A NURSERY COMPANION

A Nursery

Provided by

Iona and Peter Opie

Companion

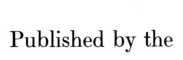

Published by the

Oxford University Press

Oxford University Press, Walton Street, Oxford OX2 6DP

London Glasgow New York Toronto
Delhi Bombay Calcutta Madras Karachi
Kuala Lumpur Singapore Hong Kong Tokyo
Nairobi Dar es Salaam Cape Town
Melbourne Wellington

and associate companies in
Beirut Berlin Ibadan Mexico City

Photographs by Derrick Witty.

ISBN 0 19 212213 4

Printed in Italy by
Amilcare Pizzi S.p.A., Milan

The reindeer, ape, lynx, racoon, tiger, elephant, lion, walrus,
hare, ass, leopard, and horse, on the two preceding pages, are from
Tommy Trip's Museum; or, A Peep at the Quadruped Race, *published*
by J. Harris and Son about 1820.

Introduction

Children have been forgiven for being children ever since adults discovered the fun of sharing a joke with them; and the person who, more than any other, brought about this amnesty was a young woman, Sarah Catherine Martin, who herself never had children, who indeed never married (she had already turned down a future king of England), and who does not even seem to have been thinking about children at the time. The score and more booklets whose contents are assembled here were the result of an innovation she made in children's literature, which to this day is not fully appreciated, let alone celebrated annually as it should be with a public holiday.

On 1 July 1805, a middle-aged publisher, John Harris, who had acquired control of the leading children's book publishing house, but had not up to this time shown great entrepreneurial ability, issued a small squarish booklet of sixteen leaves, 5 × 4 inches, described on the cover simply as *The Comic Adventures of Old Mother Hubbard, and Her Dog: Illustrated with Fifteen Elegant Engravings on Copper-plate.* Inside appeared the now-familiar metrical tale of an old woman whose cupboard was bare; and each verse was paired with a graphic but not very accomplished engraving.

The probability is that Harris himself did not know who had conceived this booklet. At the beginning of the nineteenth century moralists and educators customarily had their names on the title-pages of their works; but those who sought to entertain the public remained discreet about it. They were not held to be advancing man's spiritual and intellectual welfare like their calf-bound contemporaries, and could not therefore expect to be admired. In only one of the frolics reproduced here, for instance, was the full name given of its author; and more than a century was to pass before the identity was revealed of 'S. C. M.' or 'S. M. C.' the initials variously appearing in early editions of *Old Mother Hubbard.*

This anonymity was not however to prove a disadvantage. The booklet, which was priced at a shilling a copy and unlike anything that had been produced before for the young—if indeed it was intended for the young—was an instant success. Harris said that within a few months he had sold 10,000 copies. By the following year twenty editions had been printed. People in all walks of life and throughout the world began ordering copies. And when the poet John Wolcot ('Peter Pindar') protested at the way the public doted on the story he was referring not to the juvenile public but to the adult.

This, of course, is the point. Nothing appeared in the booklet which identified it as being for the young other than that it could be obtained in London 'at the Original Juvenile Library, the Corner of St Paul's Church-yard', where, as was well known, books for mature readers were also on sale. In fact the booklet was produced with such style, being engraved throughout and printed on good quality paper, that it was difficult to think it could be intended for the young. Nobody, however venerable, needed to feel embarrassed at being seen with it; and everybody, even the evangelical Mrs Trimmer, thought it amusing. Harris had produced a booklet, inadvertently as it now appears, which an adult could appreciate as well as a child; and although, like any publisher of the present day who has something novel on his hands, he did not understand precisely why the book was so successful, he possessed energy and ability, and immediately set about looking for other material that could be issued in the same manner.

Before the year was out he had produced *Whimsical Incidents, or the Power of Music, a Poetic Tale by a near Relation of Old Mother Hubbard;* and thereafter *A Continuation of The Comic Adventures of Old Mother Hubbard,* and *A Sequel to The Comic Adventures of Old Mother Hubbard.* But none of them had the zest of the original. In the spring of 1806, in addition to *The Adventures of Old Dame Trot,* he tried *Pug's Visit, or The Disasters of Mr Punch* in which the humour was too simian even for those days; and followed, decorously but not altogether felicitously, with *The Happy Courtship, Merry Marriage, and Pic Nic Dinner, of Cock Robin and Jenny Wren* and *The Talking Bird: or, Dame Trudge and her Parrot.* In 1807, however, he hit the jackpot again with another curiosity, a 32-line poem entitled *The Butterfly's Ball,* written—it was correctly rumoured —by a historian and member of parliament; and soon after he had the good fortune to be offered, as a sequel to *The Butterfly's Ball, The Peacock 'At Home'* by Charlotte Smith's younger sister, Mrs Dorset, which proved as successful if not more so. (The texts of both pieces are in *The Oxford Book of Children's Verse.*) He then showed he was beginning to understand the type of verse required by resurrecting Cowper's *Diverting History of John Gilpin* and Goldsmith's *Elegy on Mrs Mary Blaize*, both of which had been conceived for adult diversion; and he proved that when such pieces are brought to life pictorially the audience for them, as Caldecott was to confirm, has no lower age limit. Even if a child lacks the ability to appreciate each nuance in such entertainments, he can enjoy, what is almost as exciting, the genuineness of his parents' enjoyment of them.

The financial rewards of these publications were, in the meantime, being observed by Harris's competitors. They could almost be heard chanting the lesson:

> Old dames who are funny
> Will bring us in money;
> There's wealth astronomical
> In a gaffer who's comical.

A comparatively new publisher, Benjamin Tabart, who had an establishment in New Bond Street, was quickish off the mark with *A True History of a Little Old Woman, who found a Silver Penny* (27 May 1806), followed by the *Memoirs of the Little Man and the Little Maid* in 1807. A branch of the Quaker publishing house of Darton was even more prompt with a fine Harris-type edition of *The Death and Burial of Cock Robin*, published 31 January 1806 (the dates are those engraved on the plates of the various productions), followed by Bloomfield's tale of a credulous old woman in *The Fakenham Ghost*, 1 April 1806. Didier and Tebbett of St. James's Street, Pall Mall, published *The Whole Particulars of that Renown'd Sportsman Sam and his Gun*, 1 January 1808; and R. Dutton, in association with Harris and Tabart, offered R. S. Sharpe's ingenious new verse-tale *The Conjuror, or the Turkey and the Ring*.

The most sprightly, as also the most unlikely leaper onto the bandwagon was the philosopher William Godwin. By the end of November 1805, he and his second wife Mary Jane had set up as booksellers and stationers off Oxford Street, and had prepared their first publication on the lines of Harris's *Old Mother Hubbard* entitled *The King and Queen of Hearts: with the Rogueries of the Knave*. Further, as if to ensure that the booklet, however trivial, would achieve a footnote in literary history, the friend whom Godwin had induced to write the verses—possibly to accompany engravings already executed—was the thirty-year-old Charles Lamb. Nor was this Godwin's only attempt to emulate *Mother Hubbard*. During the next couple of years he and Mary Jane produced a series of little booklets on the same model, culminating in 1808 with *Mounseer Nongtongpaw; or, The Discoveries of John Bull in a Trip to Paris*, reproduced here; and although the ascription has not previously been made the author of this *jeu d'esprit* seems to have been Godwin's own talented ten-year-old daughter, the future Mary Shelley.

Godwin's business did not prosper; Harris's did. Godwin might express the view that 'the true object of education, like that of any other moral process, is the generation of happiness'; yet fail to appreciate that to the majority of people the vital source of happiness is not intellectual development but a cosy human relationship. From a literary point of view Godwin's publications were more worthwhile than Harris's. Some of them, such as *Swiss Family Robinson* and the Lambs' *Tales from Shakespeare*, have become landmarks; but Harris's books had, at one and the same time, a feeling of class and of kindliness that has reappeared only rarely in children's literature over the years, but is to be seen in, for instance, the publications of Felix Summerly in the 1840s, and the Puffin Picture Books in the 1940s.

In 1819 Harris, now joined by his son, sensed the way trade was going and boldly up-marketed his nursery books still further. The small squarish booklets that he had been producing, priced at 'One Shilling plain, and Eighteen-pence coloured', gave little room for verse and illustration on the same page despite the skill of his engravers in lettering. He now increased the size to 7 × 4 inches so that picture and text, which was now usually type-set, could appear together in comfort; and, more importantly, he determined that in his new 'Cabinet of Amusement and Instruction' the books should be issued only with their illustrations coloured. The soft outlines and deep shading that had been a feature of the copperplates hitherto were largely discarded. The new illustrations had sharp outlines, the minimum of shading, and, very often, almost blank backgrounds. Indeed pictures of this period which have been found uncoloured look as flat as a stage-set designed for colour television that is viewed in black and white. But when the professional colourists had been at work with their water-colours the results were wonderful. In those inflationary days the coach and hackney trade, who wanted only the best, were quite prepared to pay Harris's standard price of 1s. 6d. each for the 'most approved Novelties for the Nursery'.

Harris commenced his new series with a splendidly re-illustrated *Comic Adventures of Old Mother Hubbard* (reproduced on pp. 28–31); and followed with a good choice of such established favourites as *The History of the House that Jack Built, The History of an Apple Pie, Cock Robin,* and *A was an Archer and Shot at a Frog,* all reproduced here. But he did not only play safe. He had come to appreciate, it seems, how much his success was due to humouring the child's guardian as well as the child, and several books in the series were genuine novelties. He issued a book of pure nonsense styled *Peter Piper's Practical Principles of Plain and Perfect Pronunciation* (pp. 75–9). He published the first-ever book of limericks, *The History of Sixteen Wonderful Old Women* (pp. 66–9). And he offered the discerning the most elegant grammar ever published, *The Paths of Learning Strewed with Flowers,* as also the most agreeable aid to the use of punctuation marks, *Punctuation Personified; or Pointing Made Easy* (see pp. 46–9 and 54–7). Here again, naturally, he was imitated. The firm of Dean and Munday issued closely related poetical grammars and aids to home education (we include their immensely popular introduction to learning music, pp. 90–3); they matched the new *Mother Hubbard* with a tale that charmed the infant Ruskin, *Dame Wiggins of Lee, and her Seven Wonderful Cats* (pp. 58–61); and they consistently undersold Harris, charging a shilling each for their almost-comparable 'coloured toy books'. Carvalho of Finsbury Square was another in the trade who undersold Harris, issuing a serviceable line of standard titles at a shilling each (his *Jack and Jill* and *Juvenile Numerator* are included here, pp. 38–9 and 36–7); and Carvalho is particularly notable in having anticipated the present volume by bringing together eleven of his picture booklets into one bumper volume of 150 illustrations which he entitled *The Infant's Own Book.*

Most of all, John Marshall, whose shop in Fleet Street was not far from Harris's in St. Paul's Churchyard, was a direct competitor. His publications can only be described as lush, his style as brash. When Harris produced *The History of Sixteen Wonderful Old Women* he rushed out a rival book of limericks, *Anecdotes and Adventures of Fifteen Gentlemen* with illustrations by Robert Cruikshank (pp. 70–4). He attempted to match *The Paths of Learning Strewed with Flowers* with *The Path of Learning Strewed with Roses.* He competed with Dean and Munday in producing texts for the then-popular memory games. And he went further than Harris, producing 'presents for good children' that went outside the usual subject-matter of children's literature. He issued colour-booklets that made fun of the current foibles of the Dandies, as in later years he might have done of the Aesthetes, or the Bright Young Things, or the pert plush-haired Punks. He produced titles such as *The Dandies' Ball, The Dandy's Wedding, The Dandy's Perambulations,* and *The Dandies' Alphabet.* Exactly how these skits were regarded by the heads of families in 1819 and 1820 we do not know. We know only that the booklets sold well; and that one child who was fascinated by them was Sheridan's eleven-year-old granddaughter, Caroline, the future notorious Mrs Norton. She emulated Mary Shelley—who incidentally was to become a friend—and herself turned child-author, writing the verses of a Dandy book. *The Dandies' Rout* is not only a remarkable production for an eleven-year-old, it is one of the best of the genre, as also now one of the rarest (even Caroline's biographers do not seem to have seen a copy), and it is included here complete with Robert Cruikshank's illustrations (pp. 113–17).

Sir Arthur Bryant has styled the years 1812–22 the 'Age of Elegance'; and the sparkle of the Regency extended even to the trappings of the nursery. The children's books produced in the first quarter of the nineteenth century have an alertness and grace not achieved in any other period. The reader has the feeling from them, as is sometimes portrayed within them, of a period in which cultured parents possessed the inclination as well as the leisure to walk hand in hand with their children, and together look at the world around them in wonder. But the temper of the times was soon to change. The belief in progress, and the possible perfectability of man, was once again to take hold of men's minds; and, as always happens in periods of moral advancement, light-heartedness was to be frowned upon. Before George IV's ten-year-reign was over the quality of the nursery colour-books had declined. Indeed the market for them had contracted. Their novelty had worn off. Cheap imitations were mass-produced by less fastidious publishers; and child-workers were employed to daub on paint where formerly skilled colourists had made a livelihood. By 1832 and the no-nonsense days of the Reform Bill, the inhabitants of the nursery were once again being looked upon as little pudding-bags to be stuffed with knowledge. The wonder is, perhaps, that people had ever thought otherwise; and the suspicion cannot be avoided that, in part, nonsense had been allowed entry to top nurseries through a misapprehension.

In 1805 nursery rhymes were not widely known and beloved as they are today; and when *The Comic Adventures of Old Mother Hubbard* was published many people imagined the tale was an original composition. It so happened that the amateur who, as we now know, supplied the sketches for the verses, was well connected. Sarah Catherine Martin was the eldest daughter of Sir Henry Martin Bt., Comptroller of the Navy; and naturally her family moved in circles that included other people active in public life. In fact when her father was naval commissioner at Portsmouth, Prince

William Henry (afterwards William IV), who was stationed at Portsmouth in 1785–6, had been a frequent visitor at his house, had fallen in love with Sarah, and offered her not only his heart but, what was scarcely his to offer, his hand in marriage. Sarah, then aged seventeen, had apparently understood the impossibility of the situation. Instead of attempting a secret liaison, she informed her family. She was promptly sent away; and although William continued for a while to affirm his attachment, it is a matter of history that in time he found himself able to bestow his affections elsewhere. Sarah's feelings, on the other hand, are unrecorded. It is known only that despite her evident attractiveness she never married; and that when in 1804 she and her sister Judith were staying at Kitley in Devon, the home of John Pollexfen Bastard, M.P., her romantic days seem to have been over; the sister the M.P. was courting was Judith. The legend in the Bastard family is that Sarah, an incessant talker, used to exasperate her host, and to obtain occasional respite he encouraged her to find some way of entertaining herself on her own. *The Comic Adventures of Old Mother Hubbard*, illustrated with her sketches (the manuscript still exists), was the result.

In every age, it seems, there are people who are incapable of understanding that nonsense is not sense in disguise. When confronted with a composition of no obvious meaning they instantly suspect a hidden meaning; and *Alice in Wonderland* is found to be a veiled account of religious controversy, Winnie the Pooh becomes a phallic symbol, and the story of the Hobbit is considered allegorical. Before Sarah's booklet was submitted to John Harris for publication she thought fit to insert a dedication:

> To J—— B—— Esq.ʳ M.P. County of ——— at whose
> suggestion and at whose House these *notable* sketches
> were design'd, this *Volume* is with all suitable deference
> *Dedicated* by his Humᵇˡᵉ Servᵗ
>
> S.C.M.

This dedication had exactly the ingredients necessary to turn a little playbook into a society bestseller, and, in so doing, make a new kind of literature fashionable in the home. It not only named, cryptically, a member of parliament, it acknowledged him to be the instigator of the pretended nonsense. Who could doubt, on this evidence, that so unusual a publication was in fact a lampoon on some political figure?

Indeed we are left with only one question. When the verve of nursery literature a hundred and sixty years ago is observed, how is it that educationalists and others have kept talking about the dreariness of books for the young in the past? The answer must be that the critics have been aware only of the dreary books, and that is of course understandable. The books that commonly survive from the past are either those that were expensive at the time or were unreadable. The books that give immediate delight, and are passed on delightedly from one reader to another, are seldom the ones that are set aside for posterity. Some of the booklets reproduced here are now so scarce they are not to be found even in the largest libraries; and we must admit to having been assembling them for this volume for more than thirty years.

Our aim here is that the pictures should be as fresh as they were in the days when Darwin and Dickens were young. To this end we have not scrupled to reproduce whichever individual picture is brightest in the two or more copies, or different editions, we may possess of a particular title; and all the illustrations are reproduced the same size as the originals, except those of *The Little Man and the Little Maid* and *Mounseer Nongtongpaw*. We hope, too, that the spirit of the original booklets has been maintained. No liberties have been taken with the texts; but we have modernized spelling and punctuation where necessary, and, like the early editors themselves, have on occasion excised the superfluous stanza. Thus the contents may be examined here of a full score of 'pretty books', more such books than were to be found probably in any home at the time; and, to do justice to their variety, we have added excerpts from a further four booklets. In fact, now that this literature can be seen in all its glory it can be allowed to stand or fall in the nursery on its merit. The historical notes at the end of the volume are for the possible interest of those with, perhaps, a sherry in their hand rather than a milkshake.

Contents

A was an Archer

A a

B b

A—was an Archer,
And shot at a frog.

B—was a Butcher,
And kept a great dog.

E e

F f

E—was an Esquire,
With pride on his brow.

F—was a Farmer,
And followed the plough.

And Shot at a Frog

C c

C—was a Captain,
    All covered with lace.

D d

D—was a Drunkard,
    And had a red face.

G g

G—was a Gamester,
    And he had ill luck.

H h

H—was a Hunter,
    And hunted a buck.

J j

J—was a Joiner,
    And built up a house.

K k

K—was a King,
    And governed a mouse.

N n

N—was a Nobleman,
    Gallant and bold.

O o

O—was an Oyster-wench,
    And a sad scold.

L l

L—was a Lady,
  And had a white hand.

M m

M—was a Merchant,
  To some foreign land.

P p

P—was a Parson,
  And wore a black gown.

Q q

Q—was a Quaker,
  And would not bow down.

13

R r

S s

R—was a Robber,
  And wanted a whip.

S—was a Sailor,
  And lived in a ship.

W w

X x

W—was a Watchman,
  And guarded the door.

X—was expensive,
  And so became poor.

Tt

T—was a Tinker,
And mended a pot.

Vv

V—was a Vintner,
A very great sot.

Yy

Y—was a Youth,
Who did not love school.

Zz

Z—was a Zany,
And looked like a fool.

Nursery Novelties

A a B b

O dear! says little Allspice A,
Here comes my Brother B;

C c D d

What will our Cousin C say
Poor Duncy D to see?

E e F f

Of Eggs, says E, I've got a nest,
His Fiddle F shall tune;

G g   H h

But Grapes, says G, I like the best
Pray H, do Hand me some;

I i   K k

With Ink a Letter I will write,
And K his Kite shall fly;

L l   M m

And L will laugh by candlelight,
While M the Moon does spy.

N n   O o

Let's eat and drink, says Nodding N,
Some fruit, says Orange O;

P p   Q q

Plum Pudding P to all will send,
And Quince from Q shall go.

R r   S s

Says R, I'll Ring the bells a score,
And S, I'll Sing a Song;

T t   U u

Says T, I'll spin my Top before
My Uncle all day long.

W w   X x

Wise W, with pan of Coals,
Shall dry cross X's bed;

Y y   Z z

And Youthful Y a kiss supply
To good old Father Z.

The History of an Apple Pie Written by Z

A Apple Pie,

B Bit it,

C Cried for it,

D Danced for it,

E Eyed it,

F Fiddled for it,

G Gobbled it,

H Hid it,

I   Inspected it,

J   Jumped over it,

K   Kicked it,

L   Laughed at it,

M   Mourned for it,

N   Nodded for it,

O   Opened it,

P   Peeped into it,

Q   Quaked for it,

R   Rode for it,

S   Skipped for it,

T   Took it,

U   Upset it,

V   Viewed it,

W   Warbled for it,

X   Xerxes drew his
      sword for it,

Y   Yawned for it,

Z   Zealous that all good boys and girls
      should be acquainted with his family,
      sat down and wrote the History of it.

THE HISTORY OF
The House that Jack Built
A DIVERTING STORY

This is the house that Jack built.

This is the malt
That lay in the house that Jack built.

This is the rat
That ate the malt,
That lay in the house that Jack built.

This is the cat
That killed the rat,
That ate the malt,
That lay in the house that Jack built.

This is the dog
That worried the cat,
That killed the rat,
That ate the malt,
That lay in the house that Jack built.

This is the cow with the crumpled horn
That tossed the dog,
That worried the cat,
That killed the rat,
That ate the malt,
That lay in the house that Jack built.

This is the maiden all forlorn
That milked the cow with the crumpled horn,
That tossed the dog,
That worried the cat,
That killed the rat,
That ate the malt,
That lay in the house that Jack built.

This is the man all tattered and torn
That kissed the maiden all forlorn,
That milked the cow with the crumpled horn,
That tossed the dog,
That worried the cat,
That killed the rat,
That ate the malt,
That lay in the house that Jack built.

This is the priest all shaven and shorn
That married the man all tattered and torn,
That kissed the maiden all forlorn,
That milked the cow with the crumpled horn,
That tossed the dog,
That worried the cat,
That killed the rat,
That ate the malt,
That lay in the house that Jack built.

This is the cock that crowed in the morn
That waked the priest all shaven and shorn,
That married the man all tattered and torn,
That kissed the maiden all forlorn,
That milked the cow with the crumpled horn,
That tossed the dog,
That worried the cat,
That killed the rat,
That ate the malt,
That lay in the house that Jack built.

This is the fox that lived under the thorn
That stole the cock that crowed in the morn,
That waked the priest all shaven and shorn,
That married the man all tattered and torn,
That kissed the maiden all forlorn,
That milked the cow with the crumpled horn,
That tossed the dog,
That worried the cat,
That killed the rat,
That ate the malt,
That lay in the house that Jack built.

This is Jack with his hound and horn
That caught the fox that lived under the thorn,
That stole the cock that crowed in the morn,
That waked the priest all shaven and shorn,
That married the man all tattered and torn,
That kissed the maiden all forlorn,
That milked the cow with the crumpled horn,
That tossed the dog,
That worried the cat,
That killed the rat,
That ate the malt,
That lay in the house that Jack built.

This is the horse of a beautiful form
That carried Jack with his hound and horn,
That caught the fox that lived under the thorn,
That stole the cock that crowed in the morn,
That waked the priest all shaven and shorn,
That married the man all tattered and torn,
That kissed the maiden all forlorn,
That milked the cow with the crumpled horn,
That tossed the dog,
That worried the cat,
That killed the rat,
That ate the malt,
That lay in the house that Jack built.

This is the stable so snug and warm
That was made for the horse of a beautiful form,
That carried Jack with his hound and horn,
That caught the fox that lived under the thorn,
That stole the cock that crowed in the morn,
That waked the priest all shaven and shorn,
That married the man all tattered and torn,
That kissed the maiden all forlorn,
That milked the cow with the crumpled horn,
That tossed the dog,
That worried the cat,
That killed the rat,
That ate the malt,
That lay in the house that Jack built.

This is the boy that every morn
Swept the stable snug and warm,
That was made for the horse of a beautiful form,
That carried Jack with his hound and horn,
That caught the fox that lived under the thorn,
That stole the cock that crowed in the morn,
That waked the priest all shaven and shorn,
That married the man all tattered and torn,
That kissed the maiden all forlorn,
That milked the cow with the crumpled horn,
That tossed the dog,
That worried the cat,
That killed the rat,
That ate the malt,
That lay in the house that Jack built.

This is Sir John Barleycorn
That treated the boy that every morn
Swept the stable snug and warm,
That was made for the horse of a beautiful form,
That carried Jack with his hound and horn,
That caught the fox that lived under the thorn,
That stole the cock that crowed in the morn,
That waked the priest all shaven and shorn,
That married the man all tattered and torn,
That kissed the maiden all forlorn,
That milked the cow with the crumpled horn,
That tossed the dog,
That worried the cat,
That killed the rat,
That ate the malt,
That lay in the house that Jack built.

The Comic Adventures of Old Mother Hubbard and her Dog

Old Mother Hubbard
Went to the cupboard,
To give the poor dog a bone:
When she came there,
The cupboard was bare,
And so the poor dog had none.

She went to the baker's
To buy him some bread;
When she came back
The dog was dead!

She went to the undertaker's
To buy him a coffin;
When she came back
The dog was laughing.

She took a clean dish
    To get him some tripe;
When she came back
    He was smoking his pipe.

She went to the alehouse
    To get him some beer;
When she came back
    The dog sat in a chair.

She went to the tavern
    For white wine and red;
When she came back
    The dog stood on his head.

She went to the fruiterer's
    To buy him some fruit;
When she came back
    He was playing the flute.

She went to the tailor's
  To buy him a coat;
When she came back
  He was riding a goat.

She went to the hatter's
  To buy him a hat;
When she came back
  He was feeding her cat.

She went to the barber's
  To buy him a wig;
When she came back
  He was dancing a jig.

She went to the cobbler's
  To buy him some shoes;
When she came back
  He was reading the news.

She went to the sempstress
   To buy him some linen;
When she came back
   The dog was spinning.

She went to the hosier's
   To buy him some hose;
When she came back
   He was dressed in his clothes.

The Dame made a curtsy,
   The dog made a bow;
The Dame said, Your servant,
   The dog said, Bow-wow.

This wonderful dog
   Was Dame Hubbard's delight,
He could read, he could dance,
   He could sing, he could write;
She gave him rich dainties
   Whenever he fed,
And erected this monument
   When he was dead.

Who killed Cock Robin?
  I, said the Sparrow,
  With my bow and arrow;
And I killed Cock Robin.

———

This is the Sparrow,
With his bow and arrow.

Who saw him die?
  I, said the Fly,
  With my little eye;
And I saw him die.

———

This is the Fly
That saw him die.

Who caught his blood?
  I, said the Fish,
  With my little dish;
And I caught his blood.

———

This is the Fish
That held the dish.

Cock Robin?

Who made his shroud?
  I, said the Beetle,
  With my little needle;
And I made his shroud.

———

This is the Beetle,
With his thread and needle.

Who will dig his grave?
  I, said the Owl,
  With my spade and shovel;
And I'll dig his grave.

———

This is the Owl so brave,
That dug Cock Robin's grave.

Who will be the Parson?
  I, said the Rook,
  With my little book;
And I will be the Parson.

———

Here's Parson Rook
Reading his book.

Who will be the Clerk?
    I, said the Lark,
    If 'tis not in the dark;
And I will be the Clerk.

————

Behold how the Lark
Says Amen, like a Clerk.

Who'll carry him to the grave?
    I, said the Kite,
    If 'tis not in the night;
And I'll carry him to the grave.

————

Behold the Kite,
How he takes his flight.

Who will be chief mourner?
    I, said the Dove,
    For I mourn for my love,
And I'll be chief mourner.

————

Here's a pretty Dove,
That mourns for her love.

Who will bear the pall?
  We, said the Wren,
  Both the cock and the hen;
And we will bear the pall.

———

Here's the Wren so small,
That held Cock Robin's pall.

Who'll sing a psalm?
  I, says the Thrush,
  As she sat in a bush;
And I'll sing a psalm.

———

Here's a fine Thrush
Singing psalms in a bush.

Who'll toll the bell?
  I, says the Bull,
  Because I can pull;
So Cock Robin, farewell.

———

Then all the Birds fell
  To sighing and sobbing,
When they heard the bell toll
  For poor Cock Robin.

1, 2

3, 4

5, 6

The Juvenile Numerator

One, two,
Buckle my shoe.
Three, four,
Open the door.
Five, six,
Pick up sticks.
Seven, eight,
Lay them straight.
Nine, ten,
A fine fat hen.

7, 8

9, 10

11, 12

13, 14

Eleven, twelve,
Ring the bell.
Thirteen, fourteen,
Maids are courting.
Fifteen, sixteen,
The maid's in the kitchen.
Seventeen, eighteen,
Boys are skating.
Nineteen, twenty,
My stomach's empty,
So pray Jane
Give me my dinner.

15, 16

17, 18

19, 20

The Adventures
and Old

Jack and Jill
  Went up the hill,
To fetch a pail of water;
  Jack fell down,
  And broke his crown,
And Jill came tumbling after.

  Then up Jack got,
  And home did trot,
As fast as he could caper;
  Dame Gill did the job,
  To plaster his nob
With vinegar and brown paper.

  Then Jill came in,
  And she did grin
To see Jack's paper plaster;
  Her mother, vexed,
  Did whip her next,
For laughing at Jack's disaster.

  This made Jill pout,
  And she ran out,
And Jack did quickly follow;
  They rode dog Ball
  Till Jill did fall,
Which made Jack laugh and halloo.

# of Jack and Jill
Dame Gill

Then Dame came out
  To enquire about,
Jill said Jack made her tumble;
  Says Jack, I'll tell
  You how she fell,
Then judge if she need grumble.

Dame Gill did grin
  As she went in,
And Jill was plagued by Jack;
  Will Goat came by,
  And made Jack cry,
And knocked him on his back.

Though Jack wasn't hurt,
  He was all over dirt;
I wish you had but seen him,
  And how Jill did jump
  Towards the pump,
And pumped on him to clean him.

Which done, all three
  Went in to tea,
And put the place all right;
  Which done, they sup,
  Then take a cup,
And wish you a good night.

The Comic Adventures of Old Dame Trot and her Cat

Here you behold Dame Trot, and here
Her comic cat you see;
Each seated in an elbow chair
As snug as they can be.

Dame Trot came home one wintry night,
A shivering, starving soul,
But Puss had made a blazing fire,
And nicely trussed a fowl.

The Dame was pleased, the fowl was
dressed,
The table set in place;
The wondrous cat began to carve,
And Goody said her grace.

The cloth withdrawn, old Goody cries,
  'I wish we'd liquor too:'
Up jumped Grimalkin for some wine,
  And soon a cork she drew.

The wine got up in Pussy's head,
  She would not go to bed;
But purred and tumbled, leaped
    and danced,
  And stood upon her head.

Old Goody laughed to see the sport,
  As though her sides would crack;
When Puss, without a single word,
  Leaped on the spaniel's back.

'Ha, ha! well done!' old Trot exclaims,
  'My cat, you gallop well;'
But Spot grew surly, growled and bit,
  And down the rider fell.

Now Goody sorely was fatigued,
  Nor eyes could open keep,
So Spot, and she, and Pussy too,
  Agreed to go to sleep.

Next morning Puss got up betimes,
  The breakfast-cloth she laid;
And ere the village clock struck eight,
  The tea and toast she made.

Goody awoke and rubbed her eyes,
  And drank her cup of tea;
Amazed to see her cat behave
  With such propriety.

The breakfast ended, Trot went out
  To see old neighbour Hards;
And coming home, she found her cat
  Engaged with Spot at cards.

Another time the Dame came in,
  When Spot demurely sat
Half lathered to the ears and eyes,
  Half shaven by the cat.

Grimalkin, having shaved her friend,
  Sat down before the glass,
And washed her face, and dressed
    her hair,
  Like any modern lass.

A hat and feather then she took,
  And stuck it on aside;
And o'er a gown of crimson silk,
  A handsome tippet tied.

Just as her dress was all complete,
  In came the good old Dame;
She looked, admired, and curtsied low,
  And Pussy did the same.

Little Rhymes

THE WINDMILL

Blow, wind, blow! and go, mill, go!
That the miller may grind his corn;
  That the baker may take it,
  And into rolls make it,
And send us some hot in the morn.

THE COCK

The cock crows in the morn
  To tell us to rise,
And that he who lies late
  Will never be wise:
For heavy and stupid,
  He can't learn his book;
So, as long as he lives,
  Like a dunce he must look.

THE LITTLE DONKEY

I'm a poor little donkey,
  And work very hard;
To my sighs and fatigues
  Master pays no regard:
Yet for him would I toil,
  And do always my best,
Would he speak to me kindly,
  And give me some rest.

for Little Folks

FANNY'S LIBRARY

THE DOG TRIM

There was once a nice little dog, Trim,
Who ne'er had ill temper or whim;
He could sit up and dance,
Could run, skip, and prance.
Who would not like little dog Trim?

THE ROCKING HORSE

When Charles has done reading
 His book every day,
Then he goes with his hoop
 In the garden to play;
Or, his whip in his hand,
 Quickly mounts up across,
And then gallops away
 On his fine rocking-horse.

LITTLE PUSS

As pussy sat upon the step,
 Taking the nice fresh air,
A neighbour's little dog came by,
 Ah, pussy, are you there?
Good morning, Mistress Pussy Cat,
 Come, tell me how you do?
Quite well, I thank you, puss replied,
 Now, tell me how are you?

The Paths of Learning

Strewed with Flowers:

OR

English Grammar Illustrated

The Six VOWELS a, e, i, o, u, y

Twenty Consonants.

b c d f g h j k l m n p q r
s t v w x z

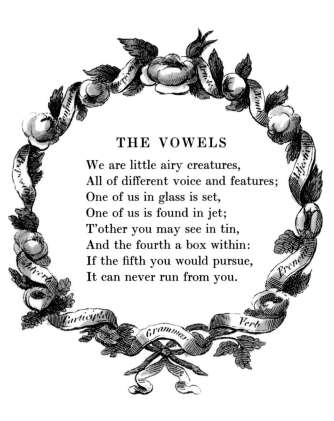

THE VOWELS

We are little airy creatures,
All of different voice and features;
One of us in glass is set,
One of us is found in jet;
T'other you may see in tin,
And the fourth a box within:
If the fifth you would pursue,
It can never run from you.

The vowels make full and perfect sounds without the help of any other letter. A consonant cannot be pronounced without the help of a vowel, nor can any word be spelt without at least one vowel.

a Rose.

a Girl.

a Cat.

VOWELS: a, e, i, o, u, y. The vowels have a perfect sound in themselves, for example a is a vowel and means one thing, as a Rose, a Cat, a Girl.

ARTICLES: a, an, the.   The article a becomes an before a vowel, and it is called an indefinite article because it means any thing, as a Cow, an Apple.

ARTICLE: the.   The is called a definite article because it means some particular thing, as I caught a pigeon, but not the pigeon with strings round its neck.

A NOUN is the name of any thing, person, or place.   Nouns mean things: whatever we can touch or see, as Gentleman, Dog, Field, Flower, Kate or Ann.

ADJECTIVE means the quality of a thing. A good girl relieving a poor lame man. Good is the quality of the girl, lame and poor of the man.

47

ADJECTIVES have three degrees of comparison. Mary is <u>short</u>—Positive degree. James is <u>shorter</u>—Comparative degree. Frank is <u>shortest</u>—Superlative degree.

PRONOUNS: <u>I</u>, <u>thou</u>, <u>he</u>, <u>she</u>, <u>my</u>, <u>thy</u>, <u>it</u>. A pronoun is a word used instead of a noun, to avoid the too frequent repetition of the same word, as Jane has a Garden and <u>she</u> waters <u>it</u>.   <u>She</u> and <u>it</u> are pronouns for Jane and Garden.

A VERB means the act of doing anything: Charles <u>rides</u> well, the horse <u>gallops</u>. The words <u>ride</u> and <u>gallop</u> are verbs.

PARTICIPLES are derived from verbs. There are two participles: first the <u>past</u>, ending in <u>ed</u>, as <u>marched</u>, <u>armed</u>; secondly the <u>present</u>, ending <u>ing</u>, as <u>walking</u>, <u>holding</u>.

ADVERBS are words <u>add</u>-ed to verbs, participles, and adjectives. Many adverbs end in <u>ly</u>. Ellen works <u>neatly</u>, sings <u>sweetly</u>, sews <u>industriously</u>.

PREPOSITIONS: <u>at</u>, <u>to</u>, <u>from</u>, <u>of</u>, <u>for</u>, <u>by</u>. Prepositions are put before nouns, verbs, and pronouns, as <u>on</u> the river <u>in</u> my boat, <u>near</u> some swans <u>with</u> my Fanny.

CONJUNCTIONS are words that join sentences and words together: you <u>and</u> I rode together <u>but</u> we did not reach London, <u>yet</u> we saw it. <u>And</u>, <u>but</u>, <u>yet</u>, are all conjunctions.

INTERJECTIONS: <u>ah</u>! <u>alas</u>! <u>O</u>! <u>la</u>! <u>fie</u>! <u>hush</u>! <u>behold</u>! Interjections are exclamations denoting any sudden emotion of the mind, either of pain, pleasure, or surprise.

The Gaping, Wide-mouthed, Waddling Frog

A gaping, wide-mouthed, waddling frog.

TWO pudding ends that won't choke a dog,
Nor a gaping, wide-mouthed, waddling frog.

THREE monkeys tied to a log,
Two pudding ends that won't choke a dog,
Nor a gaping, wide-mouthed, waddling frog.

FOUR horses stuck in a bog,
Three monkeys tied to a log,
Two pudding ends that won't choke a dog,
Nor a gaping, wide-mouthed, waddling frog.

FIVE puppies by our dog Ball,
That daily for their breakfast call;
Four horses stuck in a bog,
Three monkeys tied to a log,
Two pudding ends that won't choke a dog,
Nor a gaping, wide-mouthed, waddling frog.

SIX beetles against the wall,
Close to an old woman's apple stall;
Five puppies by our dog Ball,
That daily for their breakfast call;
Four horses stuck in a bog,
Three monkeys tied to a log,
Two pudding ends that won't choke a dog,
Nor a gaping, wide-mouthed, waddling frog.

SEVEN lobsters in a dish,
As good as any heart can wish;
Six beetles against the wall,
Close to an old woman's apple stall;
Five puppies by our dog Ball,
That daily for their breakfast call;
Four horses stuck in a bog,
Three monkeys tied to a log,
Two pudding ends that won't choke a dog,
Nor a gaping, wide-mouthed, waddling frog.

EIGHT joiners in Joiners' Hall,
Working with their tools and all;
Seven lobsters in a dish,
As good as any heart can wish;
Six beetles against the wall,
Close to an old woman's apple stall;
Five puppies by our dog Ball,
That daily for their breakfast call;
Four horses stuck in a bog,
Three monkeys tied to a log,
Two pudding ends that won't choke a dog,
Nor a gaping, wide-mouthed, waddling frog.

NINE peacocks in the air,
I wonder how they all got there,
You don't know, nor I don't care;
Eight joiners in Joiners' Hall,
Working with their tools and all;
Seven lobsters in a dish,
As good as any heart can wish;
Six beetles against the wall,
Close to an old woman's apple stall;
Five puppies by our dog Ball,
Who daily for their breakfast call;
Four horses stuck in a bog,
Three monkeys tied to a log,
Two pudding ends that won't choke a dog,
Nor a gaping, wide-mouthed, waddling frog.

TEN comets in the sky,
Some low and some high;
Nine peacocks in the air,
I wonder how they all got there,
You don't know, nor I don't care;
Eight joiners in Joiners' Hall,
Working with their tools and all;
Seven lobsters in a dish,
As good as any heart can wish;
Six beetles against the wall,

Close to an old woman's apple stall;
Five puppies by our dog Ball,
Who daily for their breakfast call;
Four horses stuck in a bog,
Three monkeys tied to a log,
Two pudding ends that won't choke a dog,
Nor a gaping, wide-mouthed, waddling frog.

ELEVEN ships sailing on the main,
Some bound for France, and some for Spain,
I wish them all safe back again;
Ten comets in the sky,
Some low and some high;
Nine peacocks in the air,
I wonder how they all got there,
You don't know, and I don't care;
Eight joiners in Joiners' Hall,
Working with their tools and all;
Seven lobsters in a dish,
As good as any heart can wish;
Six beetles against the wall,
Close to an old woman's apple stall;
Five puppies by our dog Ball,
Who daily for their breakfast call;
Four horses stuck in a bog,
Three monkeys tied to a log,
Two pudding ends that won't choke a dog,
Nor a gaping, wide-mouthed, waddling frog.

TWELVE huntsmen with horns and hounds,
Hunting over other men's grounds;
Eleven ships sailing on the main,
Some bound for France, and some for Spain,
I wish them all safe back again;
Ten comets in the sky,
Some low and some high;
Nine peacocks in the air,
I wonder how they all got there,
You don't know, and I don't care;
Eight joiners in Joiners' Hall,
Working with their tools and all;
Seven lobsters in a dish,
As good as any heart can wish;
Six beetles against a wall,
Close to an old woman's apple stall;
Five puppies by our dog Ball,
Who daily for their breakfast call;
Four horses stuck in a bog,
Three monkeys tied to a log,
Two pudding ends that won't choke a dog,
Nor a gaping, wide-mouthed, waddling frog.

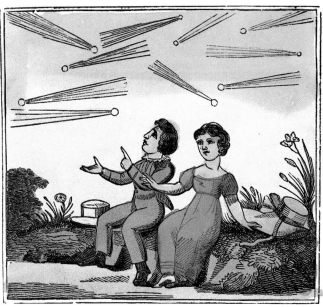

Punctuation
Personified

Young Robert could read, but he gabbled so fast,
And ran on with such speed, that all meaning
    he lost;
Till one morning he met Mr Stops by the way,
Who advised him to listen to what he should say.
Then, ent'ring the house, he a riddle repeated,
To show, *without stops*, how the ear may be cheated.

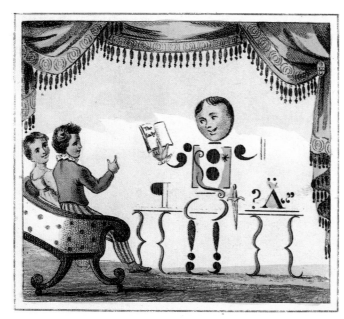

MR STOPS READING TO
ROBERT AND HIS SISTER

'Every lady in this land
Has twenty nails upon each hand
Five and twenty on hands and feet
And this is true without deceit.'
But when the stops were placed aright,
The real sense was brought to light.

COUNSELLOR COMMA, marked thus   **,**

Here counsellor COMMA the reader may view,
Who knows neither guile nor repentance;
A straightforward path he resolves to pursue
By dividing short parts of a sentence;
As 'Charles can sing, whistle, leap, tumble,
    and run'—
Yet so *brief* is each pause, that he merely counts one.

ENSIGN SEMICOLON, marked thus **;**

See, how SEMICOLON is strutting with pride!
Into two or more parts he'll a sentence divide.
As 'John's a good scholar; but George is a better.
One wrote a fair copy; the other a letter.'
Without this gay ensign we little could do;
And when he appears we must pause and count two.

A COLON, marked thus **:**

The COLON consists of two dots, as you see;
And remains within sight whilst you count one,
    two, three.
'Tis used where the sense is complete, though
    but part
Of the sentence you're reading, or learning by heart.
As 'Gold is deceitful: it bribes to destroy.'
'Young James is admired: he's a very good boy.'

A PERIOD or FULL STOP, marked thus •

The full-faced gentleman here shown
To all my friends, no doubt, is known.
In him the PERIOD we behold,
Who stands his ground whilst four are told;
And always ends the perfect sentence,
As 'Crime is followed by repentance.'

THE INTERROGATIVE POINT ?

What little crooked man is this?
He's called INTERROGATION, Miss.
He's always asking this and that,
As 'What's your name?' 'Whose dog is that?'
And for your answer, he will stay
While you 'One, two, three, four' can say.

THE EXCLAMATION POINT !

or Note of Admiration

This youth, so struck with admiration,
Is of a wondering generation,
With face so long and thin and pale,
He cries 'Oh! what a wondrous tale!'
While you count four, he stops, and then,
Admiring! he goes on again.

AN APOSTROPHE ,

The comma, placed as here you see,
From the word LOV'D has snatched a letter.
It bears the name APOSTROPHE;
And, perhaps, you can't contrive a better.
In poetry 'tis chiefly found,
Where sense should coincide with sound.

A DASH — CIRCUMFLEX ∧
BREVE ∪ DIAERESIS ··
HYPHEN — ACUTE ACCENT ╱
GRAVE ACCENT ╲ PARENTHESIS ( )

A DASH and a CIRCUMFLEX here form a hat;
A BREVE serves to mark out the face;
DIAERESIS, too, and the HYPHEN come pat,
As a breast and a neck in their place.
The arms are the ACCENTS, both GRAVE
     and ACUTE,
And for legs the PARENTHESIS nicely may suit.

A QUOTATION 66 99

Two commas standing on their heads,
Their orders are obeying;
Two others, risen from their beds,
Their best respects are paying.
These four are ushers of much use,
As they great authors introduce.

A CARET, marked thus ∧

If you a letter are inditing
And make an error in your writing,
By leaving out a word, or two,
The CARET may be used by you;
               new
As 'This∧book to Charles I send,
                    my
And hope to please∧dearest friend.'

Dame Wiggins
of Lee —

Dame Wiggins of Lee
   Was a worthy old soul,
As e'er threaded a nee-dle,
   Or washed in a bowl;
She held mice and rats
   In such antipathy,
That seven fine cats
   Kept Dame Wiggins of Lee.

The rats and mice scared
   By this fierce whiskered crew,
The poor seven cats
   Soon had nothing to do;
So, as any one idle
   She ne'er loved to see,
She sent them to school,
   Did Dame Wiggins of Lee.

But soon she grew tired
   Of living alone,
So she sent for her cats
   From school to come home;
Each rowing a wherry,
   Returning you see;
The frolic made merry
   Dame Wiggins of Lee.

— and her Seven Wonderful Cats

To give them a treat,
　　She ran out for some rice;
When she came back,
　　They were skating on ice;
'I shall soon see one down,
　　Aye, perhaps two, or three,
I'll bet half-a-crown,'
　　Said Dame Wiggins of Lee.

While, to make a nice pudding,
　　She went for a sparrow,
They were wheeling a sick lamb
　　Home, in a barrow.
'You shall all have some sprats
　　For your humanity,
My seven good cats,'
　　Said Dame Wiggins of Lee.

While she ran to the field,
　　To look for its dam,
They were warming the bed
　　For the poor sick lamb;
They turned up the clothes
　　As neat as could be;
'I shall ne'er want a nurse,'
　　Said Dame Wiggins of Lee.

She wished them good night,
    And went up to bed;
When lo! in the morning,
    The cats were all fled.
But soon, what a fuss!
    'Where can they all be?
Here, pussy, puss, puss!'
    Cried Dame Wiggins of Lee.

The Dame's heart was nigh broke,
    So she sat down to weep;
When she saw them come back,
    Each riding a sheep.
She fondled and patted
    Each purring Tom-my:
'Ah! welcome my dears,'
    Said Dame Wiggins of Lee.

The Dame was unable
    Her pleasure to smother,
To see the sick lamb
    Jump up to its mother;
In spite of the gout,
    And a pain in her knee,
She went dancing about,
    Did Dame Wiggins of Lee.

The Farmer soon heard
　　Where his sheep went astray,
And arrived at Dame's door,
　　With his faithful dog Tray.
He knocked with his crook,
　　And, the stranger to see,
Out of window did look
　　Dame Wiggins of Lee.

For their kindness he had them
　　All drawn by his team,
And gave them some field-mice,
　　And raspberry cream;
Said he, 'All my farm
　　You shall presently see,
For I honour the cats
　　Of Dame Wiggins of Lee.'

For the care of his lamb,
　　And their comical pranks,
He gave them a ham,
　　And abundance of thanks.
'I wish you good day,
　　My fine fellows,' said he;
'My compliments, pray,
　　To Dame Wiggins of Lee.'

Dame Dearlove's Ditties for the Nursery

SQUIRE FROG'S VISIT

Squire Frog he went to Mouse's hall,
　'Heigh ho!' says Brittle;
Squire Frog he went to Mouse's hall,
Dressed out quite smart for a supper and ball;
　With a namby, pamby,
　Mannikin, pannikin,
　'Heigh!' says Barnaby Brittle.

Mr Rat he bowed, and welcomed him in,
　'Heigh ho!' says Brittle;
Mr Rat he bowed, and welcomed him in,
And all the Miss Mouses went curtseying;
　With a namby, pamby,
　Mannikin, pannikin,
　'Heigh!' says Barnaby Brittle.

Then Rat did for the fiddles call,
　'Heigh ho!' says Brittle;

Then Rat did for the fiddles call,
And requested that Froggy would open the ball;
　With a namby, pamby,
　Mannikin, pannikin,
　'Heigh!' says Barnaby Brittle.

They danced until the clock struck one,
　'Heigh ho!' says Brittle;
They danced until the clock struck one,
When Rat said supper was begun;
　With a namby, pamby,
　Mannikin, pannikin,
　'Heigh!' says Barnaby Brittle.

'Good wife, pray hand that dish below,'
　'Heigh ho!' says Brittle;
'Good wife, pray hand that dish below,
For Froggy and I love harico;
　With a namby, pamby,
　Mannikin, pannikin,
　'Heigh!' says Barnaby Brittle.

'Go fetch the wine from off the hob,'
　'Heigh ho!' says Brittle;
'Go fetch the wine from off the hob,
For Froggy and I must hob or nob;'
　With a namby, pamby,
　Mannikin, pannikin,
　'Heigh!' says Barnaby Brittle.

Then Frog he begged to give a toast,
　'Heigh ho!' says Brittle;
Then Frog he begged to give a toast,
'May the French never fricassee me on our
　　coast.'
　With a namby, pamby,
　Mannikin, pannikin,
　'Heigh!' says Barnaby Brittle.

Whilst thus they merry-making sat,
　'Heigh ho!' says Brittle;
Whilst thus they merry-making sat,
Came bouncing in the great black cat;
　With a namby, pamby,
　Mannikin, pannikin,
　'Heigh!' says Barnaby Brittle.

She seized the old rat in a trice,
  'Heigh ho!' says Brittle;
She seized the old rat in a trice,
And vengeance vowed on all the mice;
  With a namby, pamby,
  Mannikin, pannikin,
  'Heigh!' says Barnaby Brittle.

When Froggy saw their dismal plight,
  'Heigh ho!' says Brittle;
When Froggy saw their dismal plight,
He thought it wiser to wish them good-night;
  With a namby, pamby,
  Mannikin, pannikin,
  'Heigh!' says Barnaby Brittle.

LITTLE PIGGY WIGGY

Come hither, little piggy wiggy,
  Come and learn your letters,
And you shall have a knife and fork
  To eat with, like your betters.
No, no! the little pig replied,
  My trough will do as well,
I'd rather eat my victuals there,
  Than go and learn to spell;
With a tingle, tangle, titmouse!
  Robin knows great A,
And B, and C, and D, and E,
  F, G, H, I, J, K.

Come hither, little pussy cat,
  If you'll your grammar study,
I'll give you silver clogs to wear,
  Whene'er the gutter's muddy.
No! whilst I grammar learn, says puss,
  Your house will in a trice
Be overrun, from top to bottom,
  With the rats and mice;
With a tingle, tangle, titmouse!
  Robin knows great A,
And B, and C, and D, and E,
  F, G, H, I, J, K.

Come hither, little puppy dog,
  I'll give you a new collar,
If you will learn to read your book,
  And be a clever scholar.
No, no! replied the puppy dog,
  I've other fish to fry,
For I must learn to guard your house,
  And bark when thieves come nigh;
With a tingle, tangle, titmouse!
  Robin knows great A,
And B, and C, and D, and E,
  F, G, H, I, J, K.

Come hither then, good little boy,
  And learn your alphabet,
And you a pair of boots and spurs,
  Like your papa's shall get.
O yes! I'll learn my alphabet,
  And when I well can read,
Perhaps papa will give me too,
  A pretty long-tailed steed;
With a tingle, tangle, titmouse!
  Robin knows great A,
And B, and C, and D, and E,
  F, G, H, I, J, K.

THE HEROES

Tweedledum and Tweedledee,
  Agreed to fight a battle;
For Tweedledum, said Tweedledee,
  Had spoiled his nice new rattle.

Just then flew down a monstrous crow,
  As black as a tar barrel;
Which frightened both those heroes so,
  They quite forgot their quarrel.

ROBIN HOOD

Pray who did kill that noble stag?
  'Twas I!—'twas I!—'twas I!
And I am called bold Robin Hood!
  Bold Robin, you must die.

Then Robin blew his bugle horn,
  And straight his archers came;
They ducked the verderer in a pool,
  And laughed to see his shame.

THE OLD WOMAN AND HER CAT

There was an old woman, who rode on a broom,
  With a high gee ho! gee humble;
And she took her Tom Cat behind for a groom
  With a bimble, bamble, bumble.

They travelled along till they came to the sky,
  With a high gee ho! gee humble;
But the journey so long made them very
      hungry,
  With a bimble, bamble, bumble.

Says Tom, 'I can find nothing here to eat,
  With a high gee ho! gee humble;
So let us go back again, I entreat,
  With a bimble, bamble, bumble.'

The old woman would not go back so soon,
  With a high gee ho! gee humble;
For she wanted to visit the man in the moon,
  With a bimble, bamble, bumble.

Says Tom, 'I'll go back by myself to our house,
  With a high gee ho! gee humble;
For there I can catch a good rat or a mouse,
  With a bimble, bamble, bumble.'

'But,' says the old woman, 'how will you go?
　　With a high gee ho! gee humble,
You shan't have my nag, I protest and vow,
　　With a bimble, bamble, bumble.'

'No, no,' says old Tom, 'I've a plan of my own,
　　With a high gee ho! gee humble.'
So he slid down the rainbow, and left her alone,
　　With a bimble, bamble, bumble.

So now if you happen to visit the sky,
　　With a high gee ho! gee humble;
And want to come back, you Tom's method
　　may try,
　　With a bimble, bamble, bumble.

The mischievous magpie flew laughing away,
　　Bumpety, bumpety, bump!
And vowed he would serve them the same the
　　next day,
　　Lumpety, lumpety, lump!

THE INDUSTRIOUS LITTLE GIRL

My mother she died, and she left me a reel,
A little silver thimble, and a pretty spinning
　　wheel;
With a high down, derry O, derry O, derry O!
High down, derry O! dance o'er the broom.

I spun all day, and I sold my yarn;
And I put in my purse all the money I did earn;
With a high down, derry O, derry O, derry O!
High down, derry O! dance o'er the broom.

And when at last I'd saved enough,
I bought me a gown of a pretty silver stuff;
With a high down, derry O, derry O, derry O!
High down, derry O! dance o'er the broom.

THE NAUGHTY MAGPIE

A farmer went trotting upon his grey mare,
　　Bumpety, bumpety, bump!
With his daughter behind him so rosy and fair;
　　Lumpety, lumpety, lump!

A magpie cried 'Caw!' and they all tumbled
　　down,
　　Bumpety, bumpety, bump!
Them are broke her knees, and the farmer his
　　crown,
　　Lumpety, lumpety, lump!

The History of Sixteen

There was an old woman named Towl,
Who went out to sea with her owl;
   But the owl was seasick,
   And screamed for physic,
Which sadly annoyed Mistress Towl.

There was an old woman at Gloucester,
Whose parrot two guineas it cost her;
   But his tongue never ceasing,
   Was vastly displeasing
To the talkative woman of Gloucester.

There was an old woman of Ealing,
She jumped till her head touched the
    ceiling,
   When 2 1 6 4
   Was announced at her door,
As a prize to the old woman of Ealing.[1]

There was an old woman at Norwich,
Who lived upon nothing but porridge;
   Parading the town,
   Made a cloak of her gown,
This thrifty old woman of Norwich.

[1] Lotteries remained a craze until abolished in 1826.

Wonderful Old Women

There was an old woman of Harrow,
Who visited in a wheelbarrow;
  And her servant before,
  Knocked loud at each door,
To announce the old woman of Harrow.

There was an old woman of Croydon,
To look young she affected the hoyden,
  And would jump and would skip
  Till she put out her hip:
Alas, poor old woman of Croydon.

There was an old woman of Gosport,
And she was one of the cross sort.
  When she dressed for the Ball
  Her wig was too small,
Which enraged this old lady of Gosport.

There was an old woman of Leith,
Who had a sad pain in her teeth;
  But the blacksmith uncouth
  Scared the pain from her tooth,
Which rejoiced the old woman of Leith.

There came an old woman from France,
Who taught grown up children to dance;
  But they were so stiff,
  She sent them home in a miff,
This sprightly old woman from France.

There was an old woman in Surrey,
Who was morn, noon, and night in a
       hurry;
  Called her husband a fool,
  Drove her children to school,
The worrying old woman of Surrey.

There dwelt an old woman at Exeter,
When visitors came it sore vexed her;
  So for fear they should eat
  She locked up all the meat,
This stingy old woman of Exeter.

There was an old woman of Bath,
And she was as thin as a lath;
  She was brown as a berry,
  With a nose like a cherry,
This skinny old woman of Bath.

68

There was an old woman in Spain,
To be civil went much 'gainst her grain;
  Yet she danced a fandango
  With General Fernando,
This whimsical woman of Spain.

There was an old woman at Leeds,
Who spent all her time in good deeds;
  She worked for the poor
  Till her fingers were sore,
This pious old woman of Leeds.

There was an old woman of Devon,
Who rose every morning at seven,
  For her house to provide,
  And to warm her inside,
This provident woman of Devon.

There lived an old woman at Lynn,
Whose nose very near touched her chin;
  You may easy suppose
  She had plenty of beaux,
This charming old woman of Lynn.

Anecdotes
and Adventures of
Fifteen Gentlemen

As a little fat man of Bombay
Was smoking one very hot day,
   A bird called a snipe
   Flew away with his pipe,
Which vexed the fat man of Bombay.

A merry old man of Oporto
Had long had the gout in his fore toe;
   And oft when he spoke
   To relate a good joke,
A terrible twinge cut it short, O!

Said a very proud farmer at Ryegate,
When the squire rode up to his high
       gate,
   'With your horse and your hound
   You had better go round,
For, I say, you shan't jump over my
       gate.'

There was a rich squire of Southwark,
From morning till night did his mouth
     work,
    So much and so fast,
    That he greatly surpassed
All Westminster, London, and
     Southwark.

There was an old captain of Dover,
Whom all the physicians gave over;
    At the sound of the drum,
    And 'The enemy's come!'
Up jumped the bold captain of
     Dover.

A butcher there was at Athlone,
Whom a beggar once asked for a bone;
    But he drove him away
    With a blow of his tray—
O! his heart was as hard as a stone.

There was a young man at St. Kitts,
Who was very much troubled with fits:
    An eclipse of the moon
    Threw him into a swoon;
Alas! poor young man of St. Kitts.

A tailor, who sailed from Quebec,
In a storm ventured once upon deck;
    But the waves of the sea
    Were as strong as could be,
And he tumbled in up to his neck.

There was an old miser at Reading,
Had a house, and a yard with a
       shed in;
    'Twas meant for a cow,
    But so small, that I vow
The poor creature could scarce get
      her head in.

There was an old soldier of Bicester
Was walking one day with his sister;
  A bull, with one poke,
  Tossed her into an oak,
Before the old gentleman missed her.

There was a sick man of Tobago
Lived long on rice-gruel and sago;
  But at last, to his bliss,
  The physician said this:
'To a roast leg of mutton you may go.'

An old gentleman living at Harwich,
At ninety was thinking of marriage:
  In came his grandson,
  Who was just twenty-one,
And went off with the bride in his
    carriage.

There was a poor man of Jamaica,
He opened a shop as a baker:
　The nice biscuits he made
　Procured him much trade,
With the little black boys of Jamaica.

A lively old man at Madeira
Thought that wine of the heart was
　　a cheerer:
　He often would say,
　'Put the bottle this way—
Absent friends! and I wish they
　　were nearer.'

There was an old merchant at Malta,
Very cross, but too stubborn to alter;
　He flew in a rage
　With poor Dr. Sage,
Who attended sick people at Malta.

Peter Piper's Practical Principles of Plain and Perfect Pronunciation

Billy Button bought a buttered
  biscuit;
Did Billy Button buy a buttered
  biscuit?
If Billy Button bought a buttered
  biscuit,
Where's the buttered biscuit Billy
  Button bought?

Captain Crackskull cracked a
  Catchpoll's Cockscomb;
Did Captain Crackskull crack a
  Catchpoll's Cockscomb?
If Captain Crackskull cracked a
  Catchpoll's Cockscomb,
Where's the Catchpoll's Cockscomb
  Captain Crackskull cracked?

Andrew Airpump asked his aunt
  her ailment;
Did Andrew Airpump ask his aunt
  her ailment?
If Andrew Airpump asked his aunt
  her ailment,
Where was the ailment of Andrew
  Airpump's aunt.

Davy Dolldrum dreamed he drove
  a dragon;
Did Davy Dolldrum dream he drove
  a dragon?
If Davy Dolldrum dreamed he drove
  a dragon,
Where's the dragon Davy Dolldrum
  dreamed he drove?

Enoch Elkrig ate an empty
    eggshell;
Did Enoch Elkrig eat an empty
    eggshell?
If Enoch Elkrig ate an empty
    eggshell,
Where's the empty eggshell
    Enoch Elkrig ate?

Francis Fribble figured on
    a Frenchman's filly;
Did Francis Fribble figure on
    a Frenchman's filly?
If Francis Fribble figured on
    a Frenchman's filly,
Where's the Frenchman's filly
    Francis Fribble figured on?

Gaffer Gilpin got a goose
    and gander;
Did Gaffer Gilpin get a goose
    and gander?
If Gaffer Gilpin got a goose
    and gander,
Where's the goose and gander
    Gaffer Gilpin got?

Humphrey Hunchback had a
    hundred hedgehogs;
Did Humphrey Hunchback have a
    hundred hedgehogs?
If Humphrey Hunchback had a
    hundred hedgehogs,
Where's the hundred hedgehogs
    Humphrey Hunchback had?

Inigo Impey itched for an
    Indian image;
Did Inigo Impey itch for an
    Indian image?
If Inigo Impey itched for an
    Indian image,
Where's the Indian image Inigo
    Impey itched for?

Lanky Lawrence lost his lass
and lobster;
Did Lanky Lawrence lose his lass
and lobster?
If Lanky Lawrence lost his lass
and lobster,
Where are the lass and lobster
Lanky Lawrence lost?

Jumping Jacky jeered a jesting
juggler;
Did Jumping Jacky jeer a jesting
juggler?
If Jumping Jacky jeered a jesting
juggler,
Where's the jesting juggler
Jumping Jacky jeered?

Matthew Mendlegs missed a
mangled monkey;
Did Matthew Mendlegs miss a
mangled monkey?
If Matthew Mendlegs missed a
mangled monkey,
Where's the mangled monkey
Matthew Mendlegs missed?

Kimbo Kemble kicked his
kinsman's kettle;
Did Kimbo Kemble kick his
kinsman's kettle?
If Kimbo Kemble kicked his
kinsman's kettle,
Where's the kinsman's kettle
Kimbo Kemble kicked?

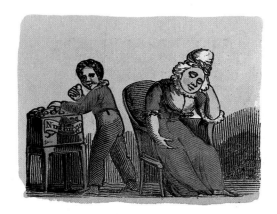

Neddy Noodle nipped his
neighbour's nutmegs;
Did Neddy Noodle nip his
neighbour's nutmegs?
If Neddy Noodle nipped his
neighbour's nutmegs,
Where are the neighbour's nutmegs
Neddy Noodle nipped?

77

Oliver Oglethorpe ogled an owl
    and oyster;
Did Oliver Oglethorpe ogle an owl
    and oyster?
If Oliver Oglethorpe ogled an owl
    and oyster,
Where are the owl and oyster
    Oliver Oglethorpe ogled?

Peter Piper picked a peck of
    pepper;
Did Peter Piper pick a peck of
    pepper?
If Peter Piper picked a peck of
    pepper,
Where's the peck of pepper
    Peter Piper picked?

Quixote Quicksight quizzed a
    queerish quidbox;
Did Quixote Quicksight quiz a
    queerish quidbox?
If Quixote Quicksight quizzed a
    queerish quidbox,
Where's the queerish quidbox
    Quixote Quicksight quizzed?

Rory Rumpus rode a raw-boned
    racer;
Did Rory Rumpus ride a raw-boned
    racer?
If Rory Rumpus rode a raw-boned
    racer,
Where's the raw-boned racer
    Rory Rumpus rode?

Sammy Smellie smelt a smell
    of smallcoal;
Did Sammy Smellie smell a smell
    of smallcoal?
If Sammy Smellie smelt a smell
    of smallcoal,
Where's the smell of smallcoal
    Sammy Smellie smelt?

Tip-Toe Tommy turned a Turk for
    twopence;
Did Tip-Toe Tommy turn a Turk for
    twopence?
If Tip-Toe Tommy turned a Turk for
    twopence,
Where's the Turk for twopence
    Tip-Toe Tommy turned?

Uncle's usher urged an ugly
    urchin;
Did Uncle's usher urge an ugly
    urchin?
If Uncle's usher urged an ugly
    urchin,
Where's the ugly urchin
    Uncle's usher urged?

Villiam Veedon viped his vig
    and vaistcoat;
Did Villiam Veedon vipe his vig
    and vaistcoat?
If Villiam Veedon viped his vig
    and vaistcoat,
Where are the vig and vaistcoat
    Villiam Veedon viped?

Walter Waddle won a walking
    wager;
Did Walter Waddle win a walking
    wager?
If Walter Waddle won a walking
    wager,
Where's the walking wager
    Walter Waddle won?

X Y Z have made my brains to
    crack-o,
X smokes, Y snuffs, and Z
    chews tobacco;
Yet oft by X Y Z much learning's
    taught;
But Peter Piper beats them all
    to nought.

The Chapter of Kings

THE ROMANS

The Romans, in England, they once
did sway

THE SAXONS

And the Saxons, they after them,
led the way,

EDMUND II AND CANUTE THE DANE

And they tugged with the Danes,
till an overthrow

HAROLD
(1066)

They both of them got by the
Norman bow.

WILLIAM I
(1066–1087)

Norman Willy, the Conqueror, long
did reign,

WILLIAM II
(1087–1100)

Red Billy, his son, by an arrow
was slain.

HENRY I
(1100–1135)

And Henry the First was a scholar
bright,

81

STEPHEN
(1135–1154)

Though Stephen was forced for
his crown to fight.

HENRY II
(1154–1189)

Second Henry, Plantagenet's name
did bear.

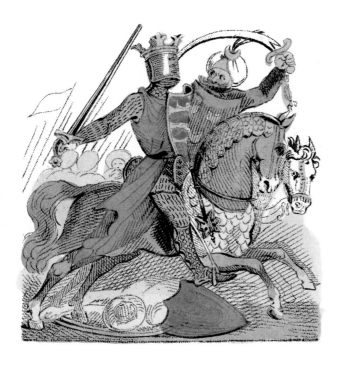

RICHARD I
(1189–1199)

Richard, Coeur de Lion, was his
son and heir.

JOHN
(1199–1216)

Famed Magna Charta we gained from
John,

HENRY III
(1216–1272)

Which Henry the Third put his seal
upon.

EDWARD I
(1272–1307)

His son, Edward the First, was a
tiger bold,

EDWARD II
(1307–1327)

Second Edward by rebels was bought
and sold;

EDWARD III
(1327–1377)

But Edward the Third was his
subjects' pride.

RICHARD II
(1377–1399)

His poor grandson, Richard, was
popped aside.

HENRY IV
(1399–1413)

Fourth Henry, of Lancaster, was a
bold wight,

HENRY V
(1413–1422)

And his son, the fifth Henry,
bravely did fight.

HENRY VI
(1422–1461)

Sixth Henry, his son, like a chick
did pout,

EDWARD IV
(1461–1483)

When fourth Edward, his cousin, had
turned him out.

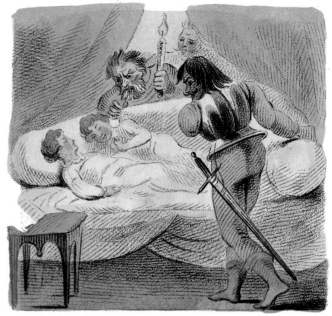

EDWARD V
(1483)

Poor Edward the Fifth, was young
killed in his bed,

RICHARD III
(1483–1485)

By his uncle, Richard, who was
knocked on the head

HENRY VII
(1485–1509)

By Henry the Seventh, who in fame
grew big,

HENRY VIII
(1509–1547)

And Henry the Eighth, who was fat
as a pig.

EDWARD VI
(1547–1553)

With Edward the Sixth we had tranquil
days,

MARY
(1553–1558)

Though Mary, his sister, made faggots
to blaze;

ELIZABETH
(1558–1603)

But good Queen Bess was a glorious
dame,

JAMES I
(1603–1625)

And King James the First from good
Scotland came.

CHARLES I
(1625–1649)

Charles the First was his son, and
a martyr made;

CHARLES II
(1660–1685)

Charles the Second, his son, was a
comical blade.

JAMES II
(1685–1688)

James the Second, his brother, when
hotly spurred,

WILLIAM AND MARY
(1689–1702)

Ran away, I assure you, from William
the Third.

ANNE
(1702–1714)

Queen Anne was victorious by land
and sea,

GEORGE I
(1714–1727)

And King George the First did with
glory sway;

GEORGE II
(1727–1760)

But as King George the Second has
long been dead,

GEORGE III
(1760–1820)

Long life to the George that we
have in his stead.

THE PRINCE REGENT

And may his son's sons to the
end of the Chapter
All come to be Kings in their turn.

The Gamut
and
Time-Table
in
Verse

Said Ann to her sister Maria, one day,
If you wish it, my dear, I will teach
    you to play;
I'll hear you your notes each day, if
    you're good,
And make them quite easy to be
    understood:
But first you'll observe, what are clear
    to be seen,
Those five straight black lines, and
    four spaces between. (1)

LINES AND SPACES.

(1)

Of the Alphabet next, seven letters you
    see,
The names of the notes, from A unto G.

And now we'll begin with first reading
    the Bass,
From G, on the first line; and A, the
    first space. (2)

The second line next, pray take notice,
  is B;
The space called the second, which
  follows, is C;
That brings you to D, which is on the
  third line,
With E, the third space, pray
  remember next time:

Then comes F, on the fourth line, and
  G, the fourth space;
If attention you pay, you'll get on
  apace;
And soon all the notes you will quite
  rightly call,
Up to A, the fifth line, and last of them
  all.

Those notes which above and below
  you discover,
Are called Ledger lines both higher
  and lower:
But at present you'll have no occasion
  to learn them,
When you have, I've no doubt, you will
  quickly discern them.

TREBLE CLEF.

E F G A B C D E F

Then the second space A, is here to
be seen,

The third line is B; C, the third space
between;

D, we find on the fourth line, as onward
we go;

And E, the fourth space, you will
presently know.

And here ends the Bass, which at
present you'll learn,

And we'll afterwards take the Treble
in turn. (3)

Here the first line is E, and F, the first
space,

And G, on the second line, next takes a
place.

With F, on the fifth line; so do not
forget

This lesson of lines and of spaces I've
set,

Which when you've repeated as well
as you're able,

We'll pass to the next rule, they call
the Time-table.

TIME-TABLE.

There's the Semibreve, longest and slowest of all, (5)
Which is equal to those two, which Minims we call, (6)
And four Crotchets here are presented to view, (7)
To equal in value those last Minims two.

Then the Eight Quavers next, which here you may see, (8)
Will with those four Crotchets exactly agree,
And of Semiquavers sixteen in number are wanted, (9)
To equal those Quavers, which last you have counted;

Then of Demisemiquavers, thirty-two in a line (10)
With the ten and six Semiquavers make even time.
Now, my dear, when this table you quite understand,
You may venture to take some new music in hand.

Peter Prim's Pride, or Proverbs

The more haste the worse speed

Every crow thinks her own young the whitest

Nobody knows where the shoe pinches so well as him that wears it

A stitch in time saves nine

We must do at Rome as the Romans do

The more the merrier, but the less the better fare

that will suit the Young or the Old

Never too old to learn

A miss is as good as a mile

The man who has no hair may lawfully wear a wig

Yawning is catching

One man's meal is another man's poison

Babes of Grace, grow apace

The Remarkable Adventures of an
Old Woman
and her Pig

AN ANCIENT TALE
IN A MODERN DRESS

A little old woman, who lived in a house
Too small for a giant, too big for a mouse,
Was sweeping her chambers (though she had
    not many),
When she found, by good fortune, a bright
    silver penny.
Delighted she seized it, and, dancing a jig,
Exclaimed, 'With this money I'll purchase a
    pig.'

So saying, away to the market she went,
And the fruits of her fortunate sweeping she
    spent
On a smooth-coated, black-spotted, curly-tailed
    thing,
Which she led off in triumph, by means of a
    string.
But how shall I paint her vexation and toil,
When, in crossing a meadow, she came to a stile,
And found neither threats nor persuasions
    would do,
To induce Mr Piggy to climb or creep through.

She coaxed him, she stroked him, she patted his
    hide,
She scolded him, threatened him, thumped him
    beside;
But coaxing, and scolding, and thumping
    proved vain,
Whilst the evening grew dark, and 'twas likely
    to rain.

The dame, out of patience, now cried in a fright
To a dog which came up, 'Pray give Piggy
    a bite,
And over the stile, sir, compel him to go,
Or here I may stay till 'tis midnight, you know.'
This request Mr Bow-wow of course must have
    heard,
But he silently stood without saying a word.
'Well, well,' said the dame, 'I'll be even with
    you,
Unkindness like this you may happen to rue.'

Then pausing, and anxiously looking around,
She saw a stout crab-stick lie flat on the ground.
'Kind stick,' she exclaimed, 'I entreat you
    to flog
This cruel, regardless, unmannerly dog,
Who will not bite Piggy, though plainly
    you see,
My pig will not stir, and there's no home
    for me.'
No reply made the stick, not a blow would it
    strike,
But crab-stick and cur remained silent alike.

'Well, this is provoking! but yonder's a fire,
And now,' said old Goody, 'I'll have my desire.'
The flame she saluted, and cried, 'Pray be quick,
Assist a poor woman, and burn this vile stick,
For 'twill not beat yon dog, though the cur will
    not bite
My pig, and I here may remain all the night.'
In vain to the flame did our sweeper appeal,
For her sufferings it would not, or perhaps *could
    not* feel.

97

An opposite element next caught her eye,
And its friendly assistance she therefore
        would try.
'Dear water,' she said, 'do extinguish this fire,
Which will not (although 'tis my ardent desire)
Consume yonder crab-stick, which, obstinate
        too,
With beating that cur will have nothing to do;
And the dog, as ill-natured, you see, as the rest,
Refuses to bite this young obstinate beast;
So here I'm compelled, most reluctant, to stay,
And here may remain till the break of the day.'
The water regardless of all that was said,
Lay perfectly still, not an effort was made.

So next to an ox her attention she turned,
And telling him how her entreaties were scorned
By the dog, by the stick, by the flame, and the
        flood,
She said, 'I beseech you, great Sir, be so good
As to drink up this water, which, everyone
        knows,
Could have put out the fire with ease, if it chose.
O grant me this favour, do pity my plight,
Or here in the fields I must stay all the night.'
The ox was unmoved, not an eye would he turn,
Though no flood would extinguish, no fire
        would burn,
No crab-stick would give Mr Bow-wow a blow,
Nor would he compel the pig forward to go.

Then kindling with rage Piggy's mistress
        cried out,
'O! here comes a man, he'll avenge me, no
        doubt.'
So once more relating her pitiful story,
She said, 'In the death of that ox I would glory.

Now therefore, good butcher, the animal kill.
I'll thank you, I'll bless you; indeed, Sir, I will.'
The butcher, however, continued his way,
Without even deigning one sentence to say.

Goody trembled with rage, yet she ventured
    to hope
A friend was at hand, when she saw a new rope.
So now with clasped hands, mournful voice, and
    bent knees,
She said, 'Hang that butcher, good rope, if you
    please;
For, though 'tis his lawful vocation each day,
An ox the barbarian refuses to slay.'
She paused for an answer; but hard was her lot,
No help, nor a word of reply could be got.

A veteran rat at this moment drew near,
And quietly stood her entreaties to hear.
So, curtseying low, 'I entreat,' said the dame,
'By your grandfather's beard, and your grand-
    mother's fame,
By the conquests your father and uncles have
    won,
And the deeds which both you and your
    brethren have done,
That your worship will not disappoint my fond
    hope,
But graciously gnaw and destroy yonder rope,
Which, spite of a moving and melting harangue,
Refuses that obstinate butcher to hang.'
But ah! in the rat no assistance was found,
And Goody's last hope seemed to fall to the
    ground.

But now kind dame Fortune at length
    interfered,
And a fierce-looking cat in a moment appeared;
A cat which was hungry, and ready to slay,
For supper, whatever might come in his way.

No sooner had, therefore, old Goody repeated
The slights with which all her petitions were
    treated,
Than Mr Grimalkin, espousing her cause,
Seized the ill-natured rat in his terrible claws.

'O spare me!' he squeaked, 'and the rope
    I'll destroy,'
But when he began his sharp teeth to employ,
The rope to hang up the cross butcher prepared;
And the butcher, that moment, most terribly
    scared,
At the head of the ox aimed a death-giving
    blow;
But submission is better than death, we all
    know;

So away, at full speed, the wise animal ran
To drink up the water. The water began
The flame to extinguish: but now 'twas the turn
Of the fire the ill-natured crab-stick to burn.
'Hold, hold!' said the stick, 'I am going to flog,
Most soundly, that obstinate cur of a dog.'
'But Sir,' said the dog, in a terrible fright,
'The old lady's pig I'm preparing to bite.'
This proved to be true, and his bite was severe.
'Oh ho!' cried the pig, 'I must not remain here,'
So over the stile he thought proper to get,
And Goody no more had occasion to fret;
For the pig to his sty was now easily led,
And she put him a trough, and clean straw for
    a bed.

Then fastened the door, and wished him good
    night.
The pig gave a grunt, as he could not speak
    right.
The old Dame went in to her neat little house,
And is now safe in bed and as snug as a mouse.

The Little Man & the Little Maid

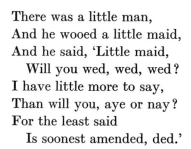

There was a little man,
And he wooed a little maid,
And he said, 'Little maid,
  Will you wed, wed, wed?
I have little more to say,
Than will you, aye or nay?
For the least said
  Is soonest amended, ded.'

The little maid replied,
'Should I be your little bride,
Pray what shall we have
  For to eat, eat, eat?
Will the flame you're only rich in,
Light a fire in the kitchen,
Or the little god of love
  Turn the spit, spit, spit?'

The little man replied,
And some say a little cried,
For his little heart was big
  With sorrow, sorrow, sorrow,
'My offers are but small,
But you have my little all,
And what we have not got
  We must borrow, borrow, borrow.'

The little man thus spoke,
His heart was almost broke,
And all for the sake
  Of her charms, charms, charms;
The little maid relents,
And, softened, she consents
The little man to take
  To her arms, arms, arms.

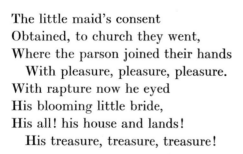

The little maid's consent
Obtained, to church they went,
Where the parson joined their hands
   With pleasure, pleasure, pleasure.
With rapture now he eyed
His blooming little bride,
His all! his house and lands!
   His treasure, treasure, treasure!

They passed their days away,
As all their neighbours say,
In feasting, mirth, and play,
   And dancing, dancing, dancing:
The little maid, they say,
Tripped merrily away,
With her little man so gay,
   Lightly prancing, prancing, prancing.

The honeymoon soon over,
No more a flaming lover,
The little man repents
   Of his folly, folly, folly;
His little cash had fled,
While he droops his pensive head,
And in sighs his sorrows vents,
   A prey to melancholy.

The little maid grew bold,
She would rant and she would scold,
And call her little man
   A great oaf, oaf, oaf.
He wished the deuce would take her,
While the butcher or the baker
Would not trust him for a chop,
   Or a loaf, loaf, loaf.

The little man reflected,
His little means neglected,
Would serve but to increase
    His sorrow, sorrow, sorrow;
To his little wife he cried,
'Let us lay our feuds aside,
And endeavour to provide
    For tomorrow, morrow, morrow.'

His little wife repented,
To his wishes she consented,
And said she could work
    With her needle, needle, needle.
The little man was not idle,
He played upon the fiddle,
And he earned a good living
    With his tweedle, tweedle, tweedle.

To the little man's great joy,
He soon had a little boy,
Which made his little heart
    Quite glad, glad, glad.
'Twas the little mother's pleasure
To nurse her little treasure,
Which rapture did impart
    To his dad, dad, dad.

Now everything was smiling,
There was no more reviling,
While cheerful plenty crowned
    Their labours, labours, labours.
The little man with joy,
Would take his little boy,
And show him all around
    To his neighbours, neighbours, neighbours.

Sam Syntax's

SIXPENCE A POTTLE, FINE STRAWBERRIES!

'Strawberries, sixpence a pottle! So nice,
That surely you will not begrudge, Sir, the price;
Of treat more delicious could epicure dream,
Than these fine large strawberries with sugar
    and cream?'

NEWS! GREAT NEWS IN THE LONDON GAZETTE!

'News! News! Here's great news in the London Gazette,
But what 'tis about, that I choose to forget—
For were I to speak all the news that befell,
I'm sure not a London Gazette could I sell!'

FLOUNDERS! JUMPING ALIVE! FINE FLOUNDERS!

'Come buy my live flounders! All jumping, ho!'
'Alive?' 'Yes, all jumping, Ma'am, two hours ago;
From sea just arrived, else may truth never thrive!
Fine flounders! Fresh flounders! all jumping alive!'

MATCHES! PLEASE TO WANT ANY MATCHES, MA'AM?

'Please want any matches, Ma'am?'—meekly and mild,
In piteous plaint ask poor woman and child;
'Do, Ma'am, buy a ha'p'orth of matches, pray do,
And blessing the poor, Ma'am, so will Heaven
    bless you.'

LAMBS TO SELL! YOUNG LAMBS TO SELL!

A flock of young lambs, each so pretty and small,
This man in his basket can carry them all,
Save one on his finger, his business to tell,
While merrily singing, 'Young lambs to sell.'

CURRANTS, RED AND WHITE, A PENNY A PINT!

'Red and white currants, your thirst to allay,
Refreshingly cool on a warm summer day.
A penny a pint! Then come taste them, and try
My red and white currants! Come buy, come buy!'

Cries of London

SWORDS, COLOURS, AND STANDARDS!

'Swords, Colours, and Standards, so brilliant
    and gay,
For juvenile heroes at Soldiers who play;
As Cornet or Ensign, to suit all conditions,
A penny each flag, and no fees for Commissions!'

SWEETBRIER AND NOSEGAYS, SO PRETTY, COME BUY!

Sweetbrier and nosegays of exquisite bloom,
So vivid their tints, and so rich their perfume,
That the well-furnished basket we're forced to admire,
And earnestly wish for a bunch of sweetbrier.

LOBSTERS! LIVE LOBSTERS!—ALL ALIVE, LOBSTERS!

'Lobsters, alive ho! Live lobsters! And dead,
Who've changed their jet armour to beautiful red,
And yet of no delicate flavour despoiled—
Lobsters alive ho! Live lobsters, and boiled!'

CHOICE BANBURY CAKES! NICE BANBURY CAKES!

'Here are Banbury cakes, so deliciously nice—
Come buy only once, and you'll shortly come twice;
Well crusted with sugar, and filled with rich fruit,
And cheap, both your pockets and palates to suit.'

WATERCRESSES!—BUY MY NICE WATERCRESSES!

Watercresses she gathers, this maiden so poor,
Our blood, if disordered, her cresses will cure;
And honest is Mary, though destined to cry,
'Fine fresh watercresses! Who'll buy them,
    who'll buy?'

DUST!—DUST HO! DUST!—DUST HO! DUST!

Now out with your dust, for the cart's at the door,
While forward goes Sam hoarsely bawling for more;
'Dust ho!' and 'Dust ho!' As he raises his voice,
His bell stuns our ears with his dissonant noise.

BUY A LIVE GOOSE!—BUY A LIVE GOOSE!

'Buy a live goose!—from my flock pick and choose,
All fat from the farm, in high order for use;
Five shillings apiece, youngand tender the whole,
As e'er flapped a wing, at the end of my pole.'

LIVE FOWLS!—LIVE FOWLS!—BUY A LIVE FOWL!

'Live fowls! Charming fowls in variety see!
Come buy them, kind customers, buy them of me.
The sluggard the cock will arouse when 'tis day,
And an egg for the weakly the pullet will lay.'

POTATOES! THREE POUNDS A PENNY, POTATOES!

Augh fait! here's a kind-hearted lass of green Erin,
Unruffled in mind, and for trifles not caring;
Who, trundling her barrow, content in her
      state is—
Still crying, 'Three pounds for a penny, potatoes!'

HOT SPICE GINGERBREAD! HOT—HOT—HOT!

For the sake of good children, young females
      and boys,
This man both his voice and his talents
      employs.
'Some fine flavoured cake,' he exclaims, 'I have got;
'Come buy my spice gingerbread, smoking hot!'

PAST TWELVE O'CLOCK, AND A CLOUDY MORNING!

While stretched on his pillow each little boy snores,
The watchman examines our windows and doors;
Or, wrapped in his coat, and the cold weather
      scorning,
Cries, 'Past twelve o'clock, and a fine starlight
      morning!'

PATROL! PATROL!

The Watch-house you here may perceive by
      the light,
And here comes the patrol to see all is right;
The watchmen on duty are used to his cry,
And while he is passing, 'All's well!' they reply.

SWEEP! SWEEP HO! SWEEP!

While quiet and warm in our beds we remain,
The Sweep trudges on, through the snow or the rain;
Though cold and half naked, the storm he defies—
Employment his object—'Sweep! Soot ho!' he cries.

PLUM PUDDING AND PIES—HOT!—PIPING HOT!

Of 'Plum pudding, all hot!' so attractive and nice,
This poor little chimney-sweep longed for a slice;
Which Tom Whimsy, the pieman, not wanting
    in sense,
Perceived, and forgave him one half the expense.

MI-EAU!—MILK BELOW, MAIDS!—MI-EAU!

That children should with milk be fed,
   My little readers know;
And this poor woman earns her bread
   By crying 'Milk below!'

ONE A PENNY, TWO A PENNY, HOT CROSS BUNS!

'Tis Good Friday morning, the little boy runs,
Along with his sister, to buy hot cross buns;
Her apron is full, yet her brother, the elf,
Unsatisfied still, must buy one for himself.

BUY A MAT, OR A HAIR BROOM!

'A mat for your door, or a brush for your room,
Come buy a stout mat, or a clean-sweeping broom;
None better than mine, I will dare to presume;
Then come, pretty maids, buy a mat or hair broom.'

CATS' MEAT, OR DOGS' MEAT!

Our faithful domestics with viands to treat,
See, here is the man with the cats' and dogs' meat.
A meat, which, it seems, is well known as he cries it,
For Pug and the cats follow Jane when she buys it.

107

Aldiboronti-
phoskyphorni-
ostikos

ALDIBORONTIPHOSKYPHORNIOSTIKOS

BOMBARDINIAN, Bashaw of three tails, who killed
Aldiborontiphoskyphorniostikos.

CHRONONHOTONTHOLOGOS, who offered a thousand
sequins for taking Bombardinian, Bashaw of three tails,
who killed Aldiborontiphoskyphorniostikos.

DICKY SNIP, the Tailor, reading the Proclamation of
Chrononhotonthologos, offering a thousand sequins for
taking Bombardinian, Bashaw of three tails, who killed
Aldiborontiphoskyphorniostikos.

ELEPHANT peeping in, as Dicky Snip the Tailor read the
Proclamation of Chrononhotonthologos, offering a thousand sequins for taking Bombardinian, Bashaw of three
tails, who killed Aldiborontiphoskyphorniostikos.

FUN! cried the 'Prentice, and pricked the Elephant's
trunk with a Needle, who peeped in as Dicky Snip the
Tailor read the Proclamation of Chrononhotonthologos,
offering a thousand sequins for taking Bombardinian,
Bashaw of three tails, who killed Aldiborontiphoskyphorniostikos.

GO IT! GO IT! cried the Elephant, and spouted mud over the 'Prentice, who pricked his trunk with a needle, as Dicky Snip the Tailor read the Proclamation of Chrononhotonthologos, offering a thousand sequins for taking Bombardinian, Bashaw of three tails, who killed Aldiborontiphoskyphorniostikos.

HA! HA! HA! laughed Hamet el Mammet, the bottle-nosed Barber of Balsora, on beholding the Elephant spout mud over the 'Prentice, who pricked his trunk with a needle, as Dicky Snip the Tailor read the Proclamation of Chrononhotonthologos, offering a thousand sequins for taking Bombardinian, Bashaw of three tails, who killed Aldiborontiphoskyphorniostikos.

ILLIKIPILLIKY! cried Snip's wife; lass a-day! 'tis too bad to titter at a body, when Hamet el Mammet, the bottle-nosed Barber of Balsora, laughed ha! ha! ha! on beholding the Elephant spout mud over the 'Prentice, who pricked his trunk with a needle, as Dicky Snip the Tailor read the Proclamation of Chrononhotonthologos, offering a thousand sequins for taking Bombardinian, Bashaw of three tails, who killed Aldiborontiphoskyphorniostikos.

KIA KHAN KREUSE, the Conjurer, transmogrified them into Pippins, because Snip's wife cried, Illikipilliky! lass a-day! 'tis too bad to titter at a body, when Hamet el Mammet, the bottle-nosed Barber of Balsora, laughed ha! ha! ha! on beholding the Elephant spout mud over the 'Prentice, who pricked his trunk with a needle, as Dicky Snip the Tailor read the Proclamation of Chrononhoton-thologos, offering a thousand sequins for taking Bom-bardinian, Bashaw of three tails, who killed Aldiboronti-phoskyphorniostikos.

LITTLE TWEEDLE gobbled them up, when Kia Khan Kreuse transmogrified them into pippins, because Snip's wife cried, Illikipilliky! lass a-day! 'tis too bad to titter at a body, when Hamet el Mammet, the bottle-nosed Barber of Balsora, laughed ha! ha! ha! on beholding the Elephant spout mud over the 'Prentice, who pricked his trunk with a needle, while Dicky Snip the Tailor read the Proclamation of Chrononhotonthologos, offering a thousand sequins for taking Bombardinian, Bashaw of three tails, who killed Aldiborontiphoskyphorniostikos.

MULEY HASSAN, Mufti of Moldavia, put on his Barnacles, to see little Tweedle gobble them up, when Kia Khan Kreuse transmogrified them into Pippins, because Snip's wife cried Illikipilliky! lass a-day! 'tis too bad to titter at a body, when Hamet el Mammet, the bottle-nosed Barber of Balsora, laughed ha! ha! ha! on beholding the Elephant spout mud over the 'Prentice, who pricked his trunk with a needle, while Dicky Snip the Tailor read the Proclamation of Chrononhotonthologos, offering a thousand sequins for taking Bombardinian, Bashaw of three tails, who killed Aldiborontiphoskyphorniostikos.

NEVER WERE SUCH TIMES! said Nicholas Hotch-Potch, as Muley Hassan, Mufti of Moldavia, put on his Barnacles to see little Tweedle gobble them up, when Kia Khan Kreuse transmogrified them into Pippins, because Snip's wife cried Illikipilliky! lass a-day! 'tis too bad to titter at a body, when Hamet el Mammet, the bottle-nosed Barber of Balsora, laughed ha! ha! ha! on beholding the Elephant spout mud over the 'Prentice, who pricked his trunk with a needle, while Dicky Snip the Tailor read the Proclamation of Chrononhotonthologos, offering a thousand sequins for taking Bombardinian, Bashaw of three tails, who killed Aldiborontiphoskyphorniostikos.

ODDS NIPPERKINS! cried Mother Bunch on her broomstick, here's a to-do! as Nicholas Hotch-Potch said, Never were such times, when Muley Hassan, Mufti of Moldavia, put on his Barnacles to see little Tweedle gobble them up, when Kia Khan Kreuse transmogrified them into Pippins, because Snip's wife said, Illikipilliky! lass a-day! 'tis too bad to titter at a body, when Hamet el Mammet, the bottle-nosed Barber of Balsora, laughed ha! ha! ha! on beholding the Elephant spout mud over the 'Prentice, who pricked his trunk with a needle, while Dicky Snip the Tailor read the Proclamation of Chrononhotonthologos, offering a thousand sequins for taking Bombardinian, Bashaw of three tails, who killed Aldiborontiphoskyphorniostikos.

PONIATOWSKY said, To jail with the Juggler and Jade, as Mother Bunch on her broomstick cried, Odds Nipperkins, here's a to-do! when Nicholas Hotch-Potch said, Never were such Times, as Muley Hassan, Mufti of Moldavia, put on his Barnacles to see little Tweedle gobble them up, when Kia Khan Kreuse transmogrified them into Pippins, because Snip's wife cried, Illikipilliky! lass a-day! 'tis too bad to titter at a body, when Hamet el Mammet, the bottle-nosed Barber of Balsora, laughed ha! ha! ha! on beholding the Elephant spout mud over the 'Prentice, who pricked his trunk with a needle, while Dicky Snip the Tailor read the Proclamation of Chrononhotonthologos, offering a thousand sequins for taking Bombardinian, Bashaw of three tails, who killed Aldiborontiphoskyphorniostikos.

QUACK! QUACK! QUACK! cried Sly Kia, and turned into a Duck, to escape from Poniatowsky, who said, To jail with the Juggler and Jade, as Mother Bunch on her broomstick cried, Odds Nipperkins, here's a to-do! when Nicholas Hotch-Potch said, Never were such Times, as Muley Hassan, Mufti of Moldavia, put on his Barnacles to see little Tweedle gobble them up, when Kia Khan Kreuse transmogrified them into Pippins, because Snip's wife said, Illikipilliky! lass a-day! 'tis too bad to titter at a body, when Hamet el Mammet, the bottle-nosed Barber of Balsora, laughed ha! ha! ha! on beholding the Elephant spout mud over the 'Prentice, who pricked his trunk with a needle, while Dicky Snip the Tailor read the Proclamation of Chrononhotonthologos, offering a thousand sequins for taking Bombardinian, Bashaw of three tails, who killed Aldiborontiphoskyphorniostikos.

RAMO SAMEE really swallowed a sword, while Sly Kia cried Quack! quack! quack! and turned into a Duck, to escape from Poniatowsky, who said, To jail with the Juggler and Jade, when Mother Bunch on her broomstick cried, Odds Nipperkins, here's a to-do! as Nicholas Hotch-Potch said, Never were such Times, as Muley Hassan, Mufti of Moldavia, put on his Barnacles to see little Tweedle gobble them up, when Kia Khan Kreuse transmogrified them into Pippins, because Snip's wife said Illikipilliky! lass a-day! 'tis too bad to titter at a body, when Hamet el Mammet, the bottle-nosed Barber of Balsora, laughed ha! ha! ha! on beholding the Elephant spout mud over the 'Prentice, who pricked his trunk with a needle, while Dicky Snip the Tailor read the Proclamation of Chrononhotonthologos, offering a thousand sequins for taking Bombardinian, Bashaw of three tails, who killed Aldiborontiphoskyphorniostikos.

SACCHARUM SWEET-TOOTH said nothing, while Ramo Samee really swallowed a sword, as Sly Kia cried Quack! quack! quack! and turned into a Duck, to escape from Poniatowsky, who said, To jail with the Juggler and Jade, when Mother Bunch on her broomstick cried, Odds Nipperkins, here's a to-do! as Nicholas Hotch-Potch said, Never were such Times, as Muley Hassan, Mufti of Moldavia, put on his Barnacles to see little Tweedle gobble them up, when Kia Khan Kreuse transmogrified them into Pippins, because Snip's wife said, Illikipilliky! lass a-day! 'tis too bad to titter at a body, when Hamet el Mammet, the bottle-nosed barber of Balsora, laughed ha! ha! ha! on beholding the Elephant spout mud over the 'Prentice, who pricked his trunk with a needle, while Dicky Snip the Tailor read the Proclamation of Chrononhotonthologos, offering a thousand sequins for taking Bombardinian, Bashaw of three tails, who killed Aldiborontiphoskyphorniostikos.

TOBY PHILPOT sat tippling with UMPO, VUMPO, and WILLY WIDEMOUTH of Wolverhampton, when X and Y, two officers, brought in the culprit, while Saccharum Sweet-tooth said nothing, though Ramo Samee really swallowed a sword, as Sly Kia cried Quack! quack! quack! and turned into a Duck, to escape from Poniatowsky, who said, To jail with the Juggler and Jade, as Mother Bunch on her Broomstick cried, Odds Nipperkins, here's a to-do! when Nicholas Hotch-Potch said, Never were such Times, as Muley Hassan, Mufti of Moldavia, put on his Barnacles to see little Tweedle gobble them up, when Kia Khan Kreuse transmogrified them into Pippins, because Snip's wife cried Illikipilliky! lass a-day! 'tis too bad to titter at a body, when Hamet el Mammet, the bottle-nosed Barber of Balsora, laughed ha! ha! ha! on beholding the Elephant spout mud over the 'Prentice, who pricked his trunk with a needle, while Dicky Snip the Tailor read the Proclamation of Chrononhotonthologos, offering a thousand sequins for taking Bombardinian, Bashaw of three tails, who killed Aldiborontiphoskyphorniostikos.

Here the spell ceased; and each in his own form scampered away to claim the reward. So off set X and Y, Willy Widemouth of Wolverhampton, Umpo, Vumpo, Toby Philpot, Saccharum Sweet-tooth, Ramo Samee, Poniatowsky, Mother Bunch on her Broomstick, Nicholas Hotch-Potch, little Tweedle, Kia Khan Kreuse, the Tailor's wife, who cried Illikipilliky, Hamet el Mammet, the bottle-nosed Barber of Balsora, the Elephant, the 'Prentice, and Dicky Snip the Tailor.

ZOROASTER's WHISKERS! exclaimed Chrononhotonthologos, here's a posse comitatus! decapitate Bombardinian, Bashaw of three tails, and divide the sequins among you. So farewell, Aldiborontiphoskyphorniostikos.

The
Dandies'
Rout

Sir Peter Tightstays was a beau
Whom everyone was proud to know.

This Dandy fine he gave a rout,
    At which all Dandies were to be;
And that not one might be left out,
    He numbered them by A, B, C.
Some foreign Dandies were there too;
A Turk, a Frenchman, and a Jew.

First comes a Dandy full of grace,
English, as you see, by his face;
His lady and three daughters, all,
Were to appear at this fine ball.

M'Carey was this Dandy's name,
Much noted in the lists of fame:
Tight stays he must put on, you know,
So calls his valet to do so.

See, in this bearded man, the Jew:
Of Dandies he had seen but few;
But seen enough of the present code
To know all dresses à-la-mode.

In boots, and calves, and whiskers
  too,
Was skilled this great and learned
  Jew;
Belthazzaar was his name, I think,
The greatest friend of Mr. Pink.

Pink washing his shirt, for 'twas
    dirty, see here,
Expecting, at night, at the ball to
    appear.
Tight trousers, short shirt, and false
    whiskers wore he,
Determined the Dandy of Dandies
    to be;
For why? a fine Dandy of fashion
    was he,
Descended from parents of high
    pedigree.

A hobby-horse apiece, they all
With speedy steps go to the ball.
Their wives and daughters there
    they'll meet,
The Dandizettes so mild and sweet,
As you will soon have an example,
In my Lady Fimple Fample,

He'd made an appointment with his
    friend the Jew,
To meet him with Dandies a
    chosen few.
And here they are, a pleasant party,
Where everyone was warm and
    hearty;

Who, in a fit of passion dire,
Threw her cap upon the fire;
Then, to show her temper sweet,
Took a stick her maid to beat,
O'erturned her table and her
    drawers,
Which, falling, made a thund'ring
    noise:

The servants coming in a hurry,
To see what caused the dreadful
    flurry.
All was soon restored to quiet,
And an end put to this fine riot.

Here's Monsieur Valeran, a great beau,
With his Turkish friend Hal Dandyso;
And here as friends they talk
    together,
About the nation and the weather.
Quoth Valeran, 'Pray take some
    wine'.
'No, thank you, Sir, I'm going to
    dine
With Simon Sprat, my ancient
    friend';
Then of his visit made an end.

Now at the ball arrived at last,
And all their salutations past,
Dancing commenced; each took
 his stand,
Ready to take his partner's hand,
Gladly an hour to beguile,
By a fair lady's lovely smile.

But now the dismal time is come,
When each must separate for his
 home.
The carriages, with noisy din,
Are now announced to those within;

And many minutes had not flown,
Till each was seated in his own,
Musing in silence on the scene
Of youth, and mirth, and sprightly
 mien,
That passed so soon from them
 away,
And will return—oh, never! nay.

Mounseer Nongtongpaw; or the Discoveries of John Bull in a Trip to Paris

John Bull, from England's happy isle,
  Too bold to dread mischance,
Resolved to leave his friends awhile,
  And take a peep at France.

He nothing knew of French indeed,
  And deemed it jabbering stuff,
For English he could write and read,
  And thought it quite enough.

Shrewd John to see, and not to prate,
  To foreign parts would roam,
That he their wonders might relate
  When snug again at home.

Arrived at Paris with his dog,
  Which he for safety muzzled,
The French flocked round him, all agog,
  And much poor John was puzzled.

He went into a tavern straight,
  Where viands smoked around,
And having gazed at every plate,
  He sat in thought profound.

He asked who gave so fine a feast,
  As fine as e'er he saw;
The landlord, shrugging at his guest,
  Said 'Je vous n'entends pas.'

'Oh! Mounseer Nongtongpaw!' said he,
  'Well, he's a wealthy man,
And seems disposed, from all I see,
  To do what good he can.

'A table set in such a style
  Holds forth a welcome sign'—
And added with an eager smile,
  'With Nongtongpaw I'll dine.'

Then to the Palais Royal on
  He trudged with honest Tray.
'Whose house is this,' said curious John,
  'So spacious and so gay?'

He rambled next to Marli's height,
    Versailles' grand scene to view,
And asked a country begging wight
    If he the master knew.

The fellow, staring, scratched his head,
    And idly stretched his jaw;
At length to John in answer said,
    'Eh! Je vous n'entends pas.'

'What, this too his!' exclaims John Bull,
    'His riches have no end.
I wish my pockets were as full—
    Would I had such a friend!'

A Frenchman, as he gaped around
    With wonder and with awe,
Salutes him with the former sound—
    'Eh! Je vous n'entends pas.'

'Hah, hah!' says John, 'Is this his place?
    Why surely he's the King—
How high is he in Fortune's grace
    Who owns so vast a thing!'

Strolling along another day,
    To feast his eager eyes,
A lady passed him, young and gay—
    He stood in fixed surprise.

Struck by her charms, he asked her name
    Of the first man he saw;
From whom, with shrugs, no answer came
    But—'Je vous n'entends pas.'

'The girl too Nongtongpaw's!' says he,
    Then cast a tender glance;
'I'm right—this Nongtongpaw must be
    The greatest man in France.'

119

A shepherd with his flock appears,
  The sheep were large and fat;
Not understanding John, he hears,
  But humbly doffs his hat.

For John with earnest looks began
  To ask whose flock he saw;
At length he heard the poor old man
  Cry—'Je vous n'entends pas.'

'Why, what the deuce!' our hero cries,
  'Are these too Nongtongpaw's?
Why surely all that meets his eyes
  He gets within his claws.'

Soon after trudged a footman nigh,
  Whose hands were full of game;
John saw them with a hungry eye
  And asked for whom they came.

But 'Je vous n'entends pas' again
  Was all that he could draw,
Which raised new wonder in his brain
  At this great Nongtongpaw.

An infant train then comes in view,
  And fills his heart with joy;
He gazes with affection true,
  And pats a smiling boy.

He asks the nurse, but asks in vain,
  Whose pretty brood appears;
For 'Je vous n'entends pas' again
  Assails his wondering ears.

Next day to view a vast balloon
    The folks come far and near,
To see it start John hurried soon,
    For every sight was dear.

He asked a woman on the ground
    Who paid for the balloon,
But 'Je vous n'entends pas' he found
    Was still the only tune.

Says he, 'I now don't wonder, Dame,
    To find 'tis his balloon,
For sure this Nongtongpaw can claim
    All that's beneath the moon.'

A splendid carriage next he sees,
    That four fine horses draw:
'Boy, say, whose coach, whose steeds are
        these?'—
    'Eh! Je vous n'entends pas.'

'Well!' honest Bull astonished roars,
    'I'm surely in a trance.
On Nongtongpaw what fortune pours—
    He must be King of France!'

Then he beheld a train of cooks,
    Whose heads rich dishes bear;
With a keen appetite he looks,
    And longs to have a share.

But 'Je vous n'entends pas' he heard
    When he the host would know.
'Aye! Nongtongpaw,' says he, ''s the word
    For all things good below.'

At last he saw a hearse pass by,
    And to the sexton said,
His bosom heaving with a sigh,
    'Pray who, my friend, is dead?'

The man the self-same answer made,
    As all had done before.
John heaved another sigh, and said,
    'Is then thy grandeur o'er?'

'I envied thee thy worldly state:
    Alas! I little knew
The malice of approaching fate—
    Poor Nongtongpaw, adieu!'

Then, pondering o'er th' untimely fall
    Of one so rich and great,
Reflections deep his mind appall
    On man's uncertain state.

For, though in manners he was rough,
    John had a feeling heart,
So thought he now had seen enough,
    And homeward should depart.

Besides, he panted to relate
    All that he heard and saw,
The pride, the pomp, the wealth, the fate,
    Of mighty Nongtongpaw.

Borne swiftly by a favouring gale,
    He reached his native ground,
And, to surprise them with the tale,
    He calls his friends around.

They hear it all with silent awe,
    Of admiration full,
And think that next to Nongtongpaw
    Is the great traveller Bull.

Notes

10–15. 'A was an Archer, and shot at a frog' filled the booklet *The Hobby-Horse, or the High Road to Learning* published by J. Harris and Son in 1820. Even at that time the alphabet was not a novelty, for rhyming alphabets of this type had been in print since the days of Queen Anne. However, its publication with a colour illustration to each line, and with the 'fat black letters' that Dickens admired, was something special; and indeed those responsible for it seem to have had a further end in view. In a manner that would be thought imprudent in a children's book today, no layer of society is shown to be without its foibles. Drunkard and squire, oyster-girl and king, are treated with uniform disrespect; and the 'High Road' the youngster is really being invited to take seems to be the one leading to recognition of man's equality.

16–19. *Nursery Novelties for Little Masters and Misses* was issued by J. Harris and Son in 1819, and was three times reprinted in the next five years. But the attraction of the booklet, then as now, probably lay in the quality of the illustrations (the hand-colouring in both copies we possess has an extraordinary radiance), and the care with which details were shown, such as the dunce's cap with protruding donkey's ears, the 'pan of coals' or warming-pan for airing the bed, the uncle's clay pipe, and the magnificent bell-ringing automaton. As in 'A was an Archer, and shot at a frog', there are only twenty-four letters since, for alphabetical purposes, the letters I and J, and U and V, were still not differentiated.

20–2. *The History of an Apple Pie, Written by Z,* was issued by J. Harris and Son with these illustrations in 1820. Clearly the history was well known at the time for parodies were rife, for instance both *The Political 'A, Apple-Pie'* and *The Constitutional Apple Pie: or Rhythmical Red Book* also appeared in 1820. Harris had originally produced an edition of the *History of an Apple Pie* in 1808; but the engravings were naïve compared with those in the booklets he was now publishing, and he did well to have it re-illustrated. The rhyme itself goes back to the seventeenth century.

23–7. *The History of the House that Jack Built,* issued by J. Harris and Son in 1820, is notable for its text as well as its illustrations. The saga is longer than usual, having a fox under the thorn, huntsman with hound and horn, horse of beautiful form, stable snug and warm, stable boy, and Sir John Barleycorn. These splendidly portrayable subjects were probably added to fill the booklet's eighteen pages. But although Harris's version remained in print for more than fifty years it never became traditional.

28–31. *The Comic Adventures of Old Mother Hubbard, and her Dog: in which is shewn the Wonderful Powers that Good Old Lady possessed in the Education of her Favorite Animal* was the grandiose title Harris gave the story of Mother Hubbard when he reissued it in 1819 to launch his 'Cabinet of Amusement and Instruction'. The verses were now presented in a smart card cover; the initials of Sarah Catherine Martin, who had been responsible for the original edition in 1805 (see Introduction), were dropped; and the fresh engravings, which are worthy of Robert Branston, were purposely designed to take colour.

32–5. *Cock Robin, A Pretty Painted Toy for either Girl or Boy; Suited to Children of All Ages,* was published by J. Harris and Son in 1819; and this date is to be seen on the coffin carried by the kite. But only the illustrations were new; 'The Funeral Song of Cock Robin' had already appeared in, for instance, *A Companion for the Nursery* printed about 1775; and, incidentally, our forerunner also included two other entertainments that appear in the present Companion: 'The House that Jack Built' and 'The Remarkable Adventures of an Old Woman and her Pig'.

36–7. *The Juvenile Numerator: or the Infant's First Step to Arithmetic* was printed and published by D. Carvalho, 74 Chiswell Street, Finsbury Square, about 1825. It was a new and much improved edition of a toy-book first produced by another publisher, G. Stevens, some fifteen years earlier.

38–9. *The Adventures of Jack and Jill and Old Dame Gill* was another booklet printed and published by D. Carvalho, 74 Chiswell Street, Finsbury Square. It appears to have been in print by about 1823, when Carvalho was selling his 'superior juvenile publications' at sixpence plain and a shilling coloured. At this time he was still only in a small way of business; but his productions seem to have sold well. Within a few years he had more than a hundred titles in print, as well as a range of twopenny books, penny books, halfpenny books, and picture sheets. Two of the engravings are signed Pickering.

40–3. *The Comic Adventures of Old Dame Trot, and her Cat: Correctly Printed from the Original in the Hubbardonian Library,* was the second volume in Harris's new 'Cabinet of Amusement and Instruction' launched in 1819. He had first produced this text, which is not the usual one, in 1806 following the success of *The Comic Adventures of Old Mother Hubbard.* The traditional verses about Dame Trot, which the other publishers were printing, were scarcely distinguishable from those about Mother Hubbard; and from Harris's point of view had nothing to commend them. Whoever sent

Harris this poetical version (and Harris seems not to have known who was the author), was the possessor of a pleasant wit; and the tale has a Puss-in-Bootsian quality that commands acceptance of the nonsense. The illustrations are clearly by the artist or engraver who did the designs for *Old Mother Hubbard* (pp. 28–31), seemingly Robert Branston.

44–5. *Little Rhymes for Little Folks*, subtitled *A Present for Fanny's Library* (later changed to *Poetry for Fanny's Library*), was published by J. Harris and Son in 1823. The rhymes are not as colourful as the pictures—we give here only the best of them—but the pictures probably always were the attraction of the book, even though the volume was almost certainly one of those the poet John Clare picked out for his children when on a visit to London in 1828. (Another was *The Remarkable Adventures of an Old Woman and Her Pig*, pp. 96–100.)

46–9. *The Paths of Learning Strewed with Flowers: or, English Grammar Illustrated* was published by J. Harris and Son in 1820. At the beginning of the booklet appeared a puff which was as wordy as such pieces often are today. 'The purpose of this little work is to obviate the reluctance children evince to the irksome and insipid task of learning the names and meaning of the component parts of grammar. Our intention is to entwine roses with instruction; and however humble our endeavours may appear, let it be recollected that the efforts of a Mouse set the Lion free from his toils.' Probably this preface was written by the publisher, since the information in the book itself is notable for its directness. The enigmatic verse on the vowels, within the wreath of flowers, has been added from John Marshall's rival publication *The Path of Learning Strewed with Roses*, 1822.

50–3. *The Gaping, Wide-mouthed, Waddling Frog* was issued by Dean and Munday in association with A. K. Newman and Co., about 1822, to provide the text for a memory game. The first player held up a thimble, or other small object, commanding the player sitting next to him, 'Take this'. The second player responded, 'What's this?' The first player said 'A gaping, wide-mouthed, waddling frog'. The second player then took it and addressed the next person in the circle with the same words, and so on until every player had inquired what it was, and had received it and passed it on. When the player who started the game had the object come back to him, he commanded his neighbour, as before, 'Take this'; but on being asked 'What's this?' he replied:

> Two pudding ends that won't choke a dog,
> Nor a gaping, wide-mouthed, waddling frog.

And so the game continued. The recitation became longer and more difficult to remember with each round; and anyone who did not repeat the sequence correctly was obliged to pay a forfeit. Such a game, naturally, received parental encouragement, since it was held to be useful in training the memory.

54–7. *Punctuation Personified: or Pointing Made Easy, By Mr. Stops*, was published by J. Harris and Son in 1824, and marked the peak of the firm's instructional productions, being not only ingenious but graceful. The case for punctuation is deftly made with the trick rhyme:

> Every lady in this land
> Has twenty nails upon each hand . . .

The illustrations are rich in felicities, as when Shakespeare's portrait represents quotation. And the counsel offered nicely combines precept with example, as in the verse on the apostrophe:

> In poetry 'tis chiefly found,
> Where sense should coincide with sound.

In fact no commendation of the work can be too high.

> Seldom are the senses pleased
> While the mind is being teased.

An American edition appeared in 1831; and a rival publication, *The Good Child's Book of Stops: or, Punctuation in Verse*, was produced by Dean and Munday with the aid of Madame Leinstein.

58–61. *Dame Wiggins of Lee, and her Seven Wonderful Cats*, which was one of the books Ruskin loved best in his childhood, was issued simultaneously by the firms of Dean and Munday and of A. K. Newman in 1823. It seems to have caught people's fancy in much the way *Old Mother Hubbard* had done eighteen years earlier, and to have appealed to such wits as Sydney Smith and R. H. Barham. Although principal credit for the tale was given to a 'lady of ninety' (reputedly a Mrs Pearson who kept a toy shop in Fleet Street opposite St Dunstan-in-the-West), the person who put it into shape is likely to have been Richard Scrafton Sharpe, a prolific but self-effacing writer of light verse, who also seems to have been responsible for *Anecdotes and Adventures of Fifteen Gentlemen* (pp. 70–4). The popularity of *Dame Wiggins* must have been due in part to the vitality of the illustrations, which have been ascribed on inconclusive evidence to R. Stennett. Stennett was certainly working for Dean and Munday at this time (see *Aldiborontiphoskyphorniostikos*, pp. 108–12), but the illustrations are not in his usual style. They may be compared however with the engravings of Dean and Munday's edition of *Old Mother Hubbard*, which Cohn gives as being engraved after designs by George Cruikshank.

62–5. *Dame Dearlove's Ditties for the Nursery; so Wonderfully Contrived, that they may be either Sung or Said by Nurse or Baby*, was issued by J. Harris and Son in 1819. The verses were a selection of those in an earlier Harris volume, *Original Ditties for the Nursery* (1805), and were now in part illustrated. Many of the ditties are remarkable for being in the metre of the old nursery rhymes, and for having something, too, of their inconsequence. In fact a few of them, including 'Tweedledum and Tweedledee', and 'A farmer went trotting upon his grey mare', are now an accepted part of the nursery canon; and if, as was claimed, the verses really were original (and the claim has not been disproved) the author was a considerable laureate of the nursery. We include here chiefly ditties that were illustrated.

66–9. *The History of Sixteen Wonderful Old Women*, published by J. Harris and Son in 1820, is notable as being the earliest book of the verses now known as limericks. In fact the verses here are the first limericks ever; that is, if the claims of the hopeful are dismissed, such as that 'Hickory dickory dock, the mouse ran up the clock' is a limerick, or Herrick's 'The Night-piece, to Julia', or Iago's drinking song in *Othello* (II. iii), or the verses concocted by the ancient Greeks at symposia, of which Aristophanes gives example at the end of *The Wasps*. Some people will have it that the limerick originated in France (the term 'limerick' seems to have gained currency only in the 1890s); others, of course, that it comes from Ireland, or, again, that it comes from France via Ireland. In *Menagiana*, vol. III, 1716, appears an epigram (noted by Boswell) on a young lady who went to a masquerade 'habillée en Jésuite', when the controversy on free will was high between the Molinistes and the Jansenists:

> On s'étonne ici que Caliste
> Ait pris l'habit de Moliniste
> Puisque cette jeune beauté
> Ote à chacun sa liberté
> N'est-ce pas une Janseniste?

Again, those who claim the limerick to be the only original verse-form in the English language, may find reassurance in the 'Ode, or Song, or both' on Jollity, featured in Christopher Smart's *The Midwife: or The Old Woman's Magazine*, vol. II, 1751:

> There was a jovial Butcher,
> He liv'd at Northern-fall-gate,
> He kept a Stall
> At Leadenhall
> And got drunk at the Boy at Aldgate.

Yet 'the fact must be faced' (as the old woman of Epping declared) that no true limerick—that is to say no example of that infectious type of verse which a hearer cannot resist trying to imitate or improve upon—has been found before 1820; while after 1820 such verses became endemic.

70–4. *Anecdotes and Adventures of Fifteen Gentlemen* is the second earliest known book of limericks, and was published by Harris's rival John Marshall probably in 1821. Its author seems to have been the versatile Richard Scrafton Sharpe, and the illustrations were later acknowledged as being by Robert Cruikshank. This is the book in which the art of the limerick was extended by the introduction of a climacteric, and sometimes far-fetched rhyme in the last line; and it is the book which—at least indirectly—brought the limerick to the Victorian nursery. Fifty years later Edward Lear acknowledged that the inspiration of his own nonsense verse was the rhyme, found here, 'There was an Old Man of Tobago'; and he seems to have been a trifle ingenuous when he implied that only this verse was known to him. He was also aquainted with the 'Old Soldier of Bicester' (two early drawings of his illustrating the verse survive); and the presumption must be that although copies of the *Fifteen Gentlemen* are today about as rare as nests in old men's beards, this was not

so in the 1830s when he himself began writing his 'nonsenses' to amuse the grandchildren of the Earl of Derby.

75–9. *Peter Piper's Practical Principles of Plain and Perfect Pronunciation* was 'Printed and Published with Pleasing Pretty Pictures, According to Act of Parliament, April 2, 1813'. Inevitably, the 'polite preface'—or puff—was in keeping with the title of the book: 'Peter Piper Puts Pen to Paper, to Produce his Peerless Performance, Proudly Presuming it will Please Princes, Peers, and Parliaments, and gain him [later changed to Procure him] the Praise and Plaudits of their Progeny and POSTERITY, Proving it Positively to be a PARAGON, or Playful, Palatable, Proverbial, Panegyrical, Philosophical, Philanthropical PHAENOMENON.' In fact the whole booklet is so stylish in its predictable lunacy that generation after generation has found it endearing. It was reprinted in its early form several times (appearing in America about 1830), and versions of it have continued to be produced to this day. When it was new, however, approval was not unanimous. A reviewer in *The London Magazine*, November 1820, dubbed it a 'vile book', and demanded to know of Harris and Son their excuse for offering it to 'the rising and risen generations'. This reaction is useful as a reminder that such a *jeu d'esprit* was still a novelty. Peter Piper should perhaps have pressed the point that his production was not intended to serve a practical purpose. The rhyme 'Peter Piper picked a peck of pickled pepper' was in fact already well known (a favourite rhyme in his youth of, for instance, Hewson Clarke, born 1787); and it had merely been thought an extension would be amusing— Holofernes not being alone in appreciating alliteration. The text we give is as amended in 1820.

80–9. *The Chapter of Kings, By Mr. Collins*, was turned into a vade-mecum for young historians by John Harris in 1818. The verses (published in the *Sporting Magazine* 1796) had originally been written as a song, sung 'with great applause' at such citadels of entertainment as the Theatre Royal, Drury Lane. But the text was then less formal; the kings were referred to as Willy, Harry, Dicky, Teddy, and the struggles for power were made light of with the refrain:

> Yet, barring all pother, the one and the other
> Were all of them Kings in their turn.

One of the earliest memories of the antiquary Edward Peacock (born 1831) was of his father singing this song for his amusement: 'I can feel now the thrill of delight the words gave me, and I trace in some degree, at least, my love for history to the impression they made on my dawning intelligence.'

'Mr. Collins', who was acknowledged but not identified on the title-page of *The Chapter of Kings*, was the eighteenth-century stage personality John Collins (*c.* 1738–1808). He made a speciality of one-man performances, giving recitations and songs of his own composing. 'The Chapter of Kings' was included in a collection of his verse called *Scripscrapologia*, a title that shows he did not take himself seriously; and he would certainly have been amazed at anyone reciting his *memoria technica* with a straight face.

89. The two illustrations in the bottom half of the page are from *Sir Harry Herald's Graphical Representation of the Dignitaries of England*, which was published by J. Harris and Son in 1820 as a 'suitable Present for the approaching Coronation'. The left-hand picture shows the King's Champion and Herald, the right-hand George IV and Queen Caroline 'in their Robes of State'. Unhappily the king and queen were never to be seen together like this, since Queen Caroline was refused entry to Westminster Abbey, and was not even granted an audience with the king. Harris had in fact been placed in a predicament similar to that which faced publishers in 1952, before the Coronation of Elizabeth II, when uncertainty prevailed whether, for constitutional reasons, the Duke of Edinburgh would ride with the Queen in the Coronation procession. Artists were obliged to picture the State Coach in such a way that it could not be seen whether or not he accompanied her. In editions of *Sir Harry Herald's Graphical Representation* issued after 1821 the picture of the king and queen together was suppressed.

90–3. *The Gamut and Time-Table in Verse*, which was advertised as being 'a familiar means of teaching Children the first Rudiments of Music', was published by Dean and Munday, in association with A. K. Newman and Co., about 1822. The book remained popular, and deservedly so, for many years. Probably numerous teachers of music found it an appropriate gift for their pupils; and, judging by copies we have seen, the book was much used for its intended purpose. Its author was a Miss C. Finch, about whom we know no more than that she subsequently wrote *The Good Child, or Sweet Home*, a volume describing 'the pleasures of school, and the amusements of the holidays'.

94–5. *Peter Prim's Pride, Or Proverbs that will Suit the Young or the Old* was published by John Harris in 1810. He had already issued *Peter Prim's Profitable Present to the Little Misses and Masters of the United Kingdom*, and other such booklets, but this new work, he said, 'as far exceeds its predecessors, as a peach exceeds a sugar plum, or as a silver medal surpasses a whipping top!' The name Peter Prim may have been traditional, being found in a ditty published in 1805:

> Peter Prim! Peter Prim!
> Why do you in stockings swim?
> Peter Prim gave this reply,
> To make such fools as you ask why!

An American edition of *Peter Prim's Pride* was published in Philadelphia as early as 1812.

96–100. *The Remarkable Adventures of an Old Woman and Her Pig* was published by the younger Harris in his 'Cabinet of Amusement and Instruction' about 1827; and it was one of the booklets the poet John Clare bought for his children when he visited London the following year. ('Tell them', Clare wrote home to his wife 'that the pictures are all colored.') Obviously he was delighted with the production; but the story itself was probably already familiar to him. The tale of a pig not jumping over a stile until bitten by a dog, which would not bite until threatened by a stick, which would not beat until approached by fire, which would not burn until about to be extinguished by water, which would not stir until about to be drunk by an ox, which would not drink until a butcher was on the point of slaughtering it, who would not trouble himself until a rope was about to hang him, was a tale that was already old, in fact it may be one of the oldest of nursery tales, for parallels exist throughout Europe. In Denmark the old woman's pig is named Fick, 'Konen och Grisen Fick'. In Sweden the story 'Gossen och Geten Näppa' centres on a refractory goat, and in France the story of 'Biquette et le loup' is little different. In Switzerland a boy goes to gather pears from a pear tree, 'Joggeli wott go Birli schüttle', and the pears refuse to fall. In Russia the story concerns a hen's efforts to obtain water for a cock that has choked on a bean. And in Germany the hen and cock version, 'Hünchen und Hänchen' was recorded in 1808 in the third volume of *Des Knaben Wunderhorn*. But the most remarkable parallel is the Hebrew chant 'Had Gadyo', which had doubtless already long been chanted when it was included in an edition of the Haggadah printed in Prague in 1590. The 'Had Gadyo' has often been given allegorical significance, as illustrating the working of Divine justice in the history of mankind, but the likelihood is that it was included in the long Passover service for the very acceptable reason that it provided the young with some light relief. A translation, issued by the Central Conference of American Rabbis, opens:

> An only kid!
> An only kid!
> My father bought
> For two zuzim.
> An only kid!
> An only kid!
>
> Then came the cat
> And ate the kid
> My father bought
> For two zuzim.
> An only kid!
> An only kid!
>
> Then came the dog
> And bit the cat
> That ate the kid
> My father bought
> For two zuzim.
> An only kid!
> An only kid!
>
> Then came a stick
> And beat the dog . . .

The closeness of the chant to the familiar nursery story becomes increasingly apparent as the chant reaches its climax:

> Then came the Holy One, Blest be He!
> And destroyed the angel of death
> That slew the butcher
> That killed the ox
> That drank the water
> That quenched the fire
> That burned the stick
> That beat the dog
> That bit the cat
> That ate the kid
> My father bought
> For two zuzim.
> An only kid!
> An only kid!

101–3. *Memoirs of the Little Man and the Little Maid: with Some Interesting Particulars of their Lives* was published by Benjamin Tabart at his Juvenile and School Library, 157 New Bond Street, in 1807. On the title-page the memoirs were stated to be 'never before published'; but later the claim was dropped as the verses had in fact been printed as long before as 1764 by Horace Walpole. What was remarkable in Tabart's booklet was the engravings which were more accomplished than any that had appeared hitherto in a publication of this kind (they are here reduced from 4 × 3½ ins), and which were the first to be sophisticated in the manner of Robert Cruikshank. The booklet was several times reprinted; and editions of it were produced in Philadelphia in 1811 and Salem, Massachusetts, in 1814.

104–7. *Sam Syntax's Description of the Cries of London, as they are Daily Exhibited in the Streets*, published by J. Harris and Son in 1820, may seem more of a novelty today than it did when it was new. Not only were street-sellers then a commonplace, but the manufacture of goods and their arrival in the home was not taken for granted as it is today. Children were not discouraged from showing curiosity about the price of goods and how they had been produced, and were taught to express gratitude for their daily fare. A host of books were produced touching upon trade; and since street-vendors were the first business people a child was likely to know, books about them were particularly popular. The cries Sam Syntax describes were amongst those most commonly heard in London in the first quarter of the nineteenth century. In the first description a 'pottle' of strawberries is, as can be seen, a measure of strawberries in a small conical basket. In the last description 'faithful domestics' refers, of course, to household pets not to domestic servants.

108–12. *Aldiborontiphoskyphorniostikos*, which was designed by the artist R. Stennett as a game for Christmas parties, was issued by Dean and Munday in November 1824. Each player had to read one division in turn as fast as he could, paying a penalty for each mistake; or, if this was felt to be too simple, all the players were to read the text at the same time but each was to start reading a word or two after the previous player had begun. Further, it was emphasized, 'Rapidity of utterance, and gesticulation, are essential'. Aldiborontiphoskyphorniostikos—a name with one letter more than the longest word in the Oxford English Dictionary—was not Stennett's invention. Henry Carey had named a wordy courtier Aldiborontiphoscophornio (*Al'-dibbo-ron'te-fos'co-for'nio*) in his burlesque *Tragedy of Chrononhotonthologos* (1734); and the fluent-tongued have delighted in it ever since. Sir Walter Scott used to call his printer James Ballantyne 'Aldiborontiphoscophornio' on account of the pompous way he spoke. A glee for three voices was composed using Carey's dialogue:

Aldiborontiphoscophornio, where left you
  Chrononhotonthologos?
Fatigued, within his tent, by the toils of war,
On downy couch reposing,
  Rigdumfunidos watching him,
While the Prince is dozing, rontiphornio,
  hotonthologos.

And a Miss Bower, who married in 1829, had the misfortune actually to be christened Aldiborontiphoscophornia. The usefulness of the republication of *Aldiborontiphoskyphorniostikos* will not, perhaps, go unnoticed by the nursery. Any adult visitor found to be overbearing can be asked innocently if he can read; and then have his ability tested with these pages.

113–17. *The Dandies' Rout*, published by John Marshall in 1820, was announced as being 'By a Young Lady of Distinction, Aged Eleven Years'. Happily the child's identity is known. She was Caroline Sheridan (born 1808), a girl so taken with becoming a writer—her grandfather was Richard Brinsley Sheridan—that on occasion she had to be forbidden pencil and paper. The story is that a friend of her parents gave her one of the 'Dandy' books that made fun of the 'exquisites' then to be seen about London, whose waists were so tight-laced they could scarcely sit down, and whose necks were stretched beyond comfort by the height of their starched collars and cravats. Probably the book Caroline was given was *The Dandies' Ball; or, High Life in the City*, which had been published the previous year; and she immediately set about writing a further tale. The manuscript was sent to John Marshall, who had it illustrated, like the others in the series, by Robert Cruikshank. Caroline's reward was to receive fifty copies, an interesting example of the recompense then thought suitable for new authors. Years later, when she had become a highly professional writer, and as the Hon. Mrs Norton was notorious in London for her vivacity and lack of discretion, a small incident occurred that she must have found flattering. She was in a bookshop in Regent Street looking for something to amuse one of her sons. The bookseller told her he had just the thing to 'please the young gentleman', and, unaware of its author, produced the colourful little booklet, *The Dandies' Rout*.

118–22. *Mounseer Nongtongpaw; or, The Discoveries of John Bull in a Trip to Paris*, was printed, early in 1808, 'for the Proprietors of the Juvenile Library, 41, Skinner Street', the proprietors of this particular 'juvenile library' or children's bookshop being William Godwin, the radical author of *Political Justice*, and his second wife Mary Jane. Charles Dibdin had recently written a song called 'Mounseer Nongtongpaw', with amusing though rather awkward verses about an English simpleton's visit to Paris:

John Bull for pastime took a prance,
Some time ago, to peep at France;
To talk of sciences and arts,
And knowledge gain'd in foreign parts.
Monsieur, obsequious, heard him speak,
And answer'd John in heathen Greek:
To all he ask'd, 'bout all he saw,
'Twas, 'Monsieur, je vous n'entends pas'.

John, to the Palais Royal come,
Its splendour almost struck him dumb,
'I say, whose house is that there here?'
'Hosse! Je vous n'entends pas, Monsieur,'
'What, Nongtongpaw again?' cries John,
'This fellow is some mighty Don:
No doubt has plenty for the maw,
I'll breakfast with this Nongtongpaw . . .'

The new version, as can be seen, followed Dibdin's tale closely, as might be expected of a young person; and in a letter Godwin wrote, 2 January 1808, sending his correspondent two manuscripts, he explained. 'That in small writing is the production of my daughter in her eleventh year, and is strictly modelled, as far as her infant talent would allow, on Dibdin's song . . .. The whole object is to keep up the joke of Nong Tong Paw being constantly taken for the greatest man in France.' Godwin's daughter, the future Mary Shelley, was as a girl both attractive and precocious. 'Her desire of knowledge is great, and her perseverance in everything she undertakes almost invincible,' wrote her father on another occasion; and the presumption must be that the verses Godwin printed were those by his daughter. She was, after all, only eighteen when she created Frankenstein and his monster. The illustrations, which are here produced from a 'plain' copy, and slightly reduced in size, were by Godwin's young friend William Mulready, the artist known today to every philatelist as the designer of the Mulready envelope.